Primal Heuristics in Integer Programming

Primal heuristics guarantee that feasible, high-quality solutions are provided at an early stage of the solving process, and thus are essential to the success of mixed-integer programming (MIP). By helping prove optimality faster, they allow MIP technology to extend to a wide variety of applications in discrete optimization.

This first comprehensive guide to the development and use of primal heuristics within MIP technology and solvers is ideal for computational mathematics graduate students and industry practitioners. Through a unified viewpoint, the book provides a unique perspective on how state-of-the-art results are integrated within the branch-and-bound approach at the core of MIP technology. It accomplishes this by highlighting all the required knowledge needed to push the heuristic side of MIP solvers to their limit and pointing out what is left to do to improve them, thus presenting heuristic approaches for MIP as part of the MIP solving process.

TIMO BERTHOLD is a lecturer at TU Berlin and a Director at FICO, leading the MIP research and development team of the FICO Xpress Solver. He is an expert on heuristic methods and computational mixed-integer linear and nonlinear programming. He has won multiple awards for his research.

ANDREA LODI is Andrew H. and Ann R. Tisch Professor at the Jacobs Technion-Cornell Institute at Cornell Tech and the Technion – IIT. His main research interests are in mixed-integer linear and nonlinear programming and data science. He has been recognized by IBM and Google faculty awards, and won the INFORMS Optimization Society 2021 Farkas Prize. He is a 2023 INFORMS Fellow.

DOMENICO SALVAGNIN is Associate Professor in Operations Research at the University of Padua, Italy. He was lead development scientist in the IBM CPLEX team in 2015–2017 and is currently a scientific consultant for FICO Xpress. His research interests include computational integer programming, constraint programming and hybrid methods for optimization.

"*Primal Heuristics in Integer Programming* by Timo Berthold, Andrea Lodi, and Domenico Salvagnin is a groundbreaking work that offers deep insights and practical approaches to designing sound heuristics for integer programming. This book will be an invaluable resource for both researchers and practitioners in the field. I will highly recommend the book to anyone looking to deepen their understanding of this crucial area."

— Matteo Fischetti, *University of Padua, Italy*

"*Primal Heuristics in Integer Programming* is a singular work in the area of computational integer optimization. It aggregates many of the most important techniques that state-of-the-art solvers use for actually producing high-quality solutions to large-scale and/or difficult mixed-integer linear optimization problems. This book fills a big gap left by the many textbooks on mixed-integer linear optimization problems. Certainly, it should be on the shelf of any student, researcher or practitioner who wants a complete picture of how such solvers work."

— Jon Lee, *University of Michigan*

Primal Heuristics in Integer Programming

TIMO BERTHOLD
FICO

ANDREA LODI
Cornell Tech

DOMENICO SALVAGNIN
University of Padua

CAMBRIDGE
UNIVERSITY PRESS

Shaftesbury Road, Cambridge CB2 8EA, United Kingdom

One Liberty Plaza, 20th Floor, New York, NY 10006, USA

477 Williamstown Road, Port Melbourne, VIC 3207, Australia

314–321, 3rd Floor, Plot 3, Splendor Forum, Jasola District Centre, New Delhi – 110025, India

103 Penang Road, #05–06/07, Visioncrest Commercial, Singapore 238467

Cambridge University Press is part of Cambridge University Press & Assessment, a department of the University of Cambridge.

We share the University's mission to contribute to society through the pursuit of education, learning and research at the highest international levels of excellence.

www.cambridge.org
Information on this title: www.cambridge.org/9781009574785

DOI: 10.1017/9781009574792

When citing this work, please include a reference to the DOI 10.1017/9781009574792

First published 2025

Cover image: Photo courtesy of Domenico Salvagnin.

A catalogue record for this publication is available from the British Library

A Cataloging-in-Publication data record for this book is available from the Library of Congress

ISBN 978-1-009-57478-5 Hardback
ISBN 978-1-009-57480-8 Paperback

Contents

Preface

This book presents primal heuristics in and for mixed-integer programming (MIP) in a unified way. Primal heuristics are a major ingredient of modern MIP solvers and a fundamental reason for their success. Although, as we will learn in the book, the work on primal heuristics has been going on pretty much from the early days of MIP, their integration within major MIP solvers has happened relatively recently, in the second part of the 2000s. Part of the reason for that is that the branch-and-cut scheme implemented in MIP solvers is inherently *dual*, that is, it relies on linear programming relaxation and its tightening by means of preprocessing, cutting planes and branching. Thus, over the years, attention has been focused on perfecting those ingredients, from both the scientific and engineering perspectives, putting on hold (if not neglecting) the *primal* side of the game.

However, when modern MIP solvers finally started to be successful in solving real-world problems routinely, the need to quickly and reliably produce high-quality feasible solutions emerged, especially from practitioners who were very keen to use the technology in industry. This is easily explained by the fact that, in practical applications, as soon as a technology is mature enough to routinely solve problems of a given size/complexity, the field naturally moves on and focuses its attention on even more challenging models (a longer horizon, a larger instance, a more integrated approach, etc.), to the point that, with notable and extremely important exceptions, often problems are actually *not* solved to proven optimality.

Of course, this (natural) shift in focus did not make the work on the dual side any less useful: there are still applications in which the optimality proof is of fundamental importance, and in any case a (tight) dual bound gives very precious information on how good the MIP solver's incumbent solution is. This is where MIP solvers with strong heuristics shine against pure heuristic methods that come without any dual information. Partially due to this incentive from the practice, over time the work on primal MIP heuristics started to focus on methods that can be swiftly integrated into complete solvers.

After the slow start before the turn of the millennium, as the urgent need for good heuristics became more and more apparent, an impressive effort, both from scientists and MIP solver developers, was put into researching primal heuristics for MIP, transforming the field very quickly and very successfully. While the old days of running MIP solvers were characterized by long waits while staring at an empty column of the logfile, hoping that it would contain the value of a primal solution at some point, nowadays, both commercial and noncommercial solvers populate such columns tremendously fast with high-quality solutions. Almost always, a first solution is found at the beginning of the root node, in many cases even before the initial linear program is solved, and by the end of the root, primal and dual bounds are often already within a few percent of each other, making the tree search, supposedly the heart of a branch-and-bound-based solver, mostly relevant for the optimality proof only. The interplay of industry demands and research interest developed, in a sense, a positive feedback loop. Practitioners required strong heuristic performance while still having knowledge of a dual bound. In turn, researchers focused on methods that can be integrated into solvers because that is where their methods would get visibility and could demonstrate relevance. As a result, the utilization of MIP solvers increased, and the practitioners' demand for even better primal heuristics increased, and so forth. As a matter of fact, black-box MIP solvers became the technology of choice to produce approximate solutions for real-world optimization problems whenever they admit a compact formulation in the MIP modeling paradigm and the application does not require instantaneous computing times.

So far, the contributions to the development described above have only been reported in scientific papers, and have – until now – not been presented in a unified way anywhere in the open literature. This book was conceived and designed precisely with the idea of giving a comprehensive overview of the field, consolidating the individual contributions into common topics, and answering the common question of how the challenging task of finding high-quality feasible solutions for hard optimization problems is practically achieved. The book is especially intended for graduate students interested in computational MIP and for practitioners routinely using MIP solvers. On the one hand, it can be used, and it has already been used in a preliminary form, for teaching an advanced master's or PhD class. On the other hand, we believe it contains all the relevant explanations and pointers to help the reader understand primal computation in MIP solvers and, hopefully, to be able to use MIP solver capabilities more efficiently.

Acknowledgments

We have been working on this book for more than half a decade. During those years, we benefited from ideas and feedback from many people, including students, colleagues and friends. In particular, we would like to thank Matteo Fischetti, for introducing two of us to the world of integer programming (in general) and to MIP heuristics (in particular): it has been a beautiful journey, Matteo, thank you! We also thank the (past and present) MIP developer teams of IBM ILOG CPLEX, FICO XPRESS and SCIP: we have all collaborated with the (very skilled) people in those teams over the last two decades on many topics, including MIP heuristics, of course, and benefited a lot from the interactions with them. Finally, we would like to thank the Cornell PhD students of the class "ORIE 6338" of Fall 2022 for proofreading some of the chapters and providing useful feedback on a preliminary version of this manuscript, and we also thank the anonymous reviewers.

<table>
<tr><td>*Timo Berthold*</td><td style="text-align:right">*Berlin*</td></tr>
<tr><td>*Andrea Lodi*</td><td style="text-align:right">*New York*</td></tr>
<tr><td>*Domenico Salvagnin*</td><td style="text-align:right">*Padua*</td></tr>
</table>

1

Introduction and Concepts

The core part of this book, Chapters 2 to 5, constitutes a comprehensive overview of different classes of MIP heuristics and describes individual heuristics in detail. In Chapter 6, we present a computational study that analyzes the impact of primal heuristics from different angles and presents two recent and innovative lines of research that extend and develop the field in promising new directions: primal heuristics for mixed-integer nonlinear programming (Chapter 7) and machine learning for primal heuristics (Chapter 8).

Before exploring these topics, this chapter introduces the main concepts and notation that we will use throughout the book. In Section 1.1, we formally introduce mixed-integer programs, while in Section 1.2 we give an overview of the different types of complete algorithms used for solving them, from the ubiquitous LP-based branch and bound to (less mainstream) primal methods. Then, in Section 1.3, we provide an overview of how primal heuristics are used inside MIP solvers: what their (typical) impact on the solving process is, and how they are scheduled within the branch-and-bound search. Then, we give a first categorization of the different types of primal heuristics inside a modern MIP solver. In Section 1.4, we introduce bound tightening and constraint propagation, essential building blocks in the design of (many) primal heuristics. In Section 1.5, we review the general concepts that are used when designing a heuristic algorithm for (general) combinatorial optimization problems, as those are relevant for the specific MIP case as well. Finally, we introduce additional concepts that are used when analyzing primal heuristics for MIP (Section 1.6) and for measuring their impact on the solving process (Section 1.7), most notably the primal integral.

1.1 Notation

Every maths book needs a section that is an accumulation of definitions. So here is ours, introducing some core concepts of computational MIP.

Definition 1.1 (MIP problem)   Let $m, n \in \mathbb{Z}_{\geq 0}$. Given a matrix $A \in \mathbb{R}^{m \times n}$, a right-hand-side vector $b \in \mathbb{R}^m$, an objective function vector $c \in \mathbb{R}^n$, a lower and an upper bound vector $l \in (\mathbb{R} \cup \{-\infty\})^n$, $u \in (\mathbb{R} \cup \{+\infty\})^n$, and a subset $\mathcal{I} \subseteq \mathcal{N} = \{1, \ldots, n\}$, the corresponding (linear) MIP problem is given by

$$
\begin{aligned}
\min \quad & c^\mathsf{T} x \\
\text{s.t.} \quad & Ax \leq b, \\
& l_j \leq x_j \leq u_j, \quad \text{for all } j \in \mathcal{N}, \\
& x_j \in \mathbb{R}, \quad \text{for all } j \in \mathcal{N} \setminus \mathcal{I}, \\
& x_j \in \mathbb{Z}, \quad \text{for all } j \in \mathcal{I}.
\end{aligned}
\tag{1.1}
$$

Note that the format given in Definition 1.1 is very general. First, maximization problems can be transformed into minimization problems by multiplying all objective function coefficients by -1. Similarly, "$\geq$" constraints can be multiplied by -1 to obtain "$\leq$" constraints. Equations can be replaced by two opposite inequalities. We assume, without loss of generality, that $l_j \leq u_j$ for all $j \in \mathcal{N}$ and $l_j, u_j \in \mathbb{Z}$ for all $j \in \mathcal{I}$.

If $\mathcal{I} = \emptyset$, problem (1.1) is called a *linear programming* (LP) problem.

For a given MIP P, the set

$$
\tilde{\mathcal{X}}(P) := \{x \in \mathbb{R}^n \mid Ax \leq b, \quad x \in [l, u], x_j \in \mathbb{Z} \text{ for all } j \in \mathcal{I}\},
$$

is called the set of *feasible solutions* of the MIP. Let $c^\star \in \mathbb{R} \cup \pm\infty$ with

$$
c^\star := \inf\{c^\mathsf{T} x \mid x \in \tilde{\mathcal{X}}\}.
$$

If $c^\star = -\infty$, we call P *unbounded*; if $c^\star = +\infty$, we call it *infeasible*. If $c^\star$ is finite, we call it the *optimal solution value* of P. A solution $x \in \tilde{\mathcal{X}}$ is called an *optimal solution* if and only if $c^\mathsf{T} x = c^\star$. If $c = 0$, we call P a *feasibility problem*.

One of the most effective techniques in MIP is the use of relaxations to provide proven lower bounds on the optimal solution value of a given problem instance. The LP relaxation of a MIP arises by omitting the integrality constraints.

Definition 1.2 (LP relaxation)   Given a MIP P of the form (1.1), the LP

$$
\begin{aligned}
\min \quad & c^\mathsf{T} x \\
\text{s.t.} \quad & Ax \leqslant b, \\
& l_j \leqslant x_j \leqslant u_j, \quad \text{for all } j \in \mathcal{N}, \\
& x_j \in \mathbb{R}, \qquad\quad \text{for all } j \in \mathcal{N},
\end{aligned}
\tag{1.2}
$$

is called the *LP relaxation of P*.

The feasible region of the LP relaxation is a polyhedron. There is always an optimal solution of the LP relaxation that is attained in a vertex of this polyhedron. Throughout this book, we will typically refer to LP-feasible solutions as $\bar{x} \in \mathbb{R}^n$, while we refer to integer feasible solutions as $\tilde{x} \in \mathbb{R}^{n-|I|} \times \mathbb{Z}^{|I|}$. For a given linear constraint

$$
\sum_{j \in \mathcal{N}} a_{ij} x_j \leq b_i,
$$

and a given point $\bar{x}$, we call the value

$$
\sum_{j \in \mathcal{N}} a_{ij} \bar{x}_j
$$

the *activity* of the constraint i with respect to the point $\bar{x}$.

By knowing the optimal objective function values of a relaxation and of a feasible (integer) solution, we get a *dual bound* and a *primal bound*, respectively, for the optimal solution value of a MIP. To measure the quality of these bounds with respect to either the optimal solution value or each other, we use the notion of gap functions.

Definition 1.3 (primal gap)   Let $\tilde{x}$ be a solution for a MIP, and let $x^\star$ be an optimal (or best known) solution for that MIP. We define the *primal gap* $\gamma^p \in [0, 1]$ of $\tilde{x}$ as

$$
\gamma^p(\tilde{x}) := \begin{cases} 0, & \text{if } c^\mathsf{T} x^\star = c^\mathsf{T} \tilde{x} = 0, \\ 1, & \text{if } c^\mathsf{T} x^\star \cdot c^\mathsf{T} \tilde{x} < 0, \\ \dfrac{|c^\mathsf{T} x^\star - c^\mathsf{T} \tilde{x}|}{\max\{|c^\mathsf{T} x^\star|,\, |c^\mathsf{T} \tilde{x}|\}}, & \text{otherwise.} \end{cases}
$$

The *primal-dual gap* is a typical measure given by MIP solvers during runtime. It is often referred to as an *optimality gap*, a name that might be considered slightly misleading since it does not explicitly describe the gap of any of the bounds to optimality, but is, rather, a worst-case estimation.

Definition 1.4 (primal-dual gap)   Let $\bar{x}$ be an optimal solution of a relaxation of a MIP and $\tilde{x}$ be a feasible solution for that MIP. We define the *primal-dual gap* $\gamma^{pd} \in \mathbb{R}_{\geq 0}$ of $\bar{x}$ and $\tilde{x}$ as

$$\gamma^{pd}(\bar{x}, \tilde{x}) := \begin{cases} 0, & \text{if } c^{\mathsf{T}}\bar{x} = c^{\mathsf{T}}\tilde{x} = 0, \\ \frac{c^{\mathsf{T}}\tilde{x} - c^{\mathsf{T}}\bar{x}}{|c^{\mathsf{T}}\bar{x}|}, & \text{if } c^{\mathsf{T}}\bar{x} \cdot c^{\mathsf{T}}\tilde{x} > 0, \\ \infty, & \text{otherwise.} \end{cases}$$

Inequalities only constrain variables against shifting their values into one direction. For the design of heuristics, it can be of interest to know into which direction a variable "is more constrained." This motivates Definition 1.5.

Definition 1.5 (variable locks)   Let a MIP in the form (1.1) be given.

(i)   The number of negative coefficients $\underline{\kappa}_j := |\{i: A_{ij} < 0\}|$ of a column $A_{.j}$ is called the number of *down-locks* of the variable x_j.

(ii)   The number of positive coefficients $\overline{\kappa}_j := |\{i: A_{ij} > 0\}|$ of a column $A_{.j}$ is called the number of *up-locks* of the variable x_j.

Further, we call a variable x_j *trivially down-roundable*, if $\underline{\kappa}_j = 0$, and hence all coefficients of the corresponding column of matrix A are nonnegative. We call a variable x_j *trivially up-roundable*, if $\overline{\kappa}_j = 0$, and hence all coefficients of the corresponding column of matrix A are nonpositive.

1.2 Algorithms for Mixed-Integer Programming

Since MIP is an $\mathcal{NP}$-hard problem,[1] all known algorithms for general MIP have a worst-case exponential runtime. *Branch and bound* [108] is the most widely used algorithm to solve mixed-integer programs. State-of-the-art MIP solvers such as SCIP [2], FICO XPRESS [60], GUROBI [85] and IBM ILOG CPLEX [101] all use *LP-based* branch and bound as a basic algorithm that is enhanced by various sophisticated subroutines to make the solvers efficient in practice.

In many cases, primal heuristics for MIP have been studied as subroutines of LP-based branch and bound; see in particular [18]. They might either be used globally, before the main part of the search, or locally for individual subproblems. The basic ideas of branch and bound are described in Section 1.2.1.

---

[1]   0–1 integer programming was among the 21 problems of Karp [103] for whom $\mathcal{NP}$-completeness was first proven. The $\mathcal{NP}$-completeness proof for the general integer case followed four years later [40].

Branch and bound, however, is not the only algorithm used to solve MIPs. A group of algorithms that is closely connected to primal heuristics are primal (exact) methods and are discussed in Section 1.2.2.

1.2.1 LP-Based Branch and Bound

> **input** : A MIP P
> **output:** An optimal solution $x^\star$ for P and the corresponding objective value $c^\star$
>
> 1 $\mathcal{L} \leftarrow \{P\}, \quad U \leftarrow \infty, \quad \tilde{x} \leftarrow \textbf{NULL}$
> 2 **if** $\mathcal{L} = \emptyset$ **then**
> 3   **return** $(\tilde{x}, U)$
> 4 **end**
> 5 Select $P_i \in \mathcal{L}, \quad \mathcal{L} \leftarrow \mathcal{L} \backslash \{P_i\}$
> 6 Solve LP relaxation of $P_i, \quad \bar{x} \leftarrow LP(P_i), \quad L_{\mathrm{loc}} \leftarrow c(\bar{x})$
>
>   /* Bounding   */
> 7 **if** $L_{loc} \geqslant U$ **then**
> 8   **goto** line 2
> 9 **end**
> 10 **if** $\bar{x} \in P$ **then**
> 11 $\tilde{x} \leftarrow \bar{x}, \quad U \leftarrow L_{\mathrm{loc}}$
> 12   **goto** line 2
> 13 **end**
> 14 Select $j \in I: \bar{x}_j \notin \mathbb{Z}$
>
>   /* Branching   */
> 15 Split P_i into $P_{2i+1} := P_i \cup \{x_j \leqslant \lfloor \bar{x}_j \rfloor\}$
> 16 $P_{2i+2} := P_i \cup \{x_j \geqslant \lceil \bar{x}_j \rceil\}, \quad \mathcal{L} \leftarrow \mathcal{L} \cup \{P_{2i+1}, P_{2i+2}\}$
> 17 **goto** line 2

Algorithm 1   LP-based branch-and-bound algorithm to solve MIPs.

The idea of branch and bound is simple, yet effective: an optimization problem is recursively split into smaller subproblems, thereby creating a search tree and implicitly enumerating all potential assignments of the integer variables. The principle algorithm is given in Algorithm 1; a visualization is given in Figure 1.1.

The task of *branching* (line 15 in Algorithm 1.1) is to successively divide the given problem instance into smaller subproblems until the individual

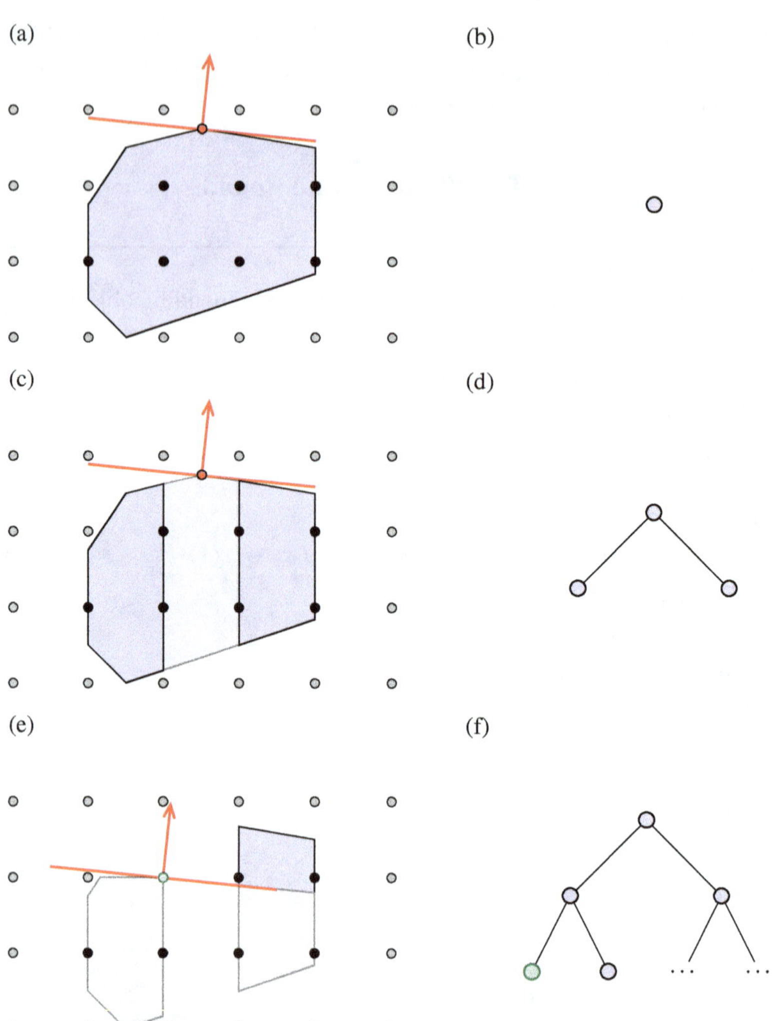

Figure 1.1 LP-based branch and bound: graphical representation. (a) The original MIP instance. (b) The root node represents the original problem. (c) Branching creates two subproblems. (d) One node for each subproblem. (e) Bounding cuts off suboptimal part. (f) The leaf node (green) is solved; the others will be recursively branched on.

subproblems are easy to solve. The best of all solutions found in the subproblems yields the global optimum. During the course of the algorithm a *branching tree* is created, with each node representing one of the subproblems. Compare Figures 1.1(a)–1.1(d). The original MIP is represented by the root node of the branching tree. Splitting the domain of one of the integer variables by rounding the fractional LP solution up and down and installing the rounded values as new lower and upper bound, respectively, creates two disjoint subproblems. Those subproblems together contain all feasible integer solutions, but the minimum of its two LP relaxations will be strictly better than the LP relaxation of the original MIP. In the branching tree, the new problems are represented as children of the root node.

The intention of *bounding* (line 7 in Algorithm 1.1) is to avoid the complete enumeration of all potential integer assignments for the initial problem, which are usually exponentially many. If a subproblem's lower (dual) bound is greater than or equal to the global upper (primal) bound, that subproblem can be pruned. Lower bounds are calculated with the help of a relaxation, which is expected to be easy to solve. Upper bounds are found if the solution of the relaxation is also (integer) feasible for the corresponding subproblem. This is visualized in Figure 1.1(e).

Most commonly, the LP relaxation (1.2) is used for dual bounding. For MIPs, the LP relaxation is simply constructed by dropping the integrality conditions; see Definition 1.2.

Various techniques have been developed to improve this basic algorithm. Besides involved strategies for making good branching and subproblem selections, this includes supplementary procedures that help in tightening the lower and upper bounds. At each subproblem, domain propagation (see Section 1.4.2) can be performed to exclude values from the variables' domains.[2] The relaxation may be strengthened by adding further valid constraints (typically linear inequalities), which cut off the optimal solution of the relaxation, but retain all feasible solutions of the MIP. In the case where a subproblem is found to be infeasible, conflict analysis might be performed to learn additional valid constraints. Primal heuristics are used as supplementary methods to improve the upper bound. A good overview on the state of the art in computational mixed-integer linear programming can be found in [1, 116].

1.2.2 Primal Methods

A possible categorization of algorithms to solve optimization problems is to subdivide them into *primal* and *dual methods*. Loosely speaking, a primal method is an algorithm that produces a sequence of feasible, suboptimal solutions until it meets a criterion proving that the current incumbent solution is optimal. By contrast, a dual method is an algorithm that produces a sequence of infeasible, "super-optimal" solutions until it finds a first feasible point –

---

[2] The domain propagation applied at each subproblem is only one of the actions that are applied to strengthen the formulation. In particular, an aggressive strengthening is performed in the root node of the MIP so as to influence as much as possible the solving process. The actions in that phase are referred to as *preprocessing* (or *presolving*) [154].

which will be an optimum. As examples, consider the primal and dual simplex algorithms.

The added advantage of LP-based branch and bound is that it produces two sequences during the course of the algorithm, providing dual and primal bounds at the same time.

By the above classification, the *cutting plane method* [79, 80] by Gomory, is a dual method; it approaches the set of feasible solutions "from the outside," solving a sequence of relaxations. As soon as the relaxation finds a point that is feasible for the MIP, the proof of optimality comes "for free."

Since the mid 1990s, there has been a rising interest in using primal methods to solve MIPs. The principal ideas of primal methods, however, date back to the 60s and 70s. *Test set algorithms* and *integral basis methods* are two important groups of primal methods for (mixed-) integer programming. Both procedures require a known feasible solution as a starting point.

Test set algorithms are motivated by the Ford–Fulkerson [67] algorithm to compute maximum flows through a network. A *test set* for a given integer program P is a finite set of n-dimensional integral vectors $\mathcal{T} \subset \mathbb{Z}^n$, such that $c^\mathsf{T} t < 0$ for all $t \in \mathcal{T}$ and that for every nonoptimal feasible solution $\tilde{x}$ of P, there exists a $t \in \mathcal{T}$ such that $\tilde{x} + t$ is a feasible solution of P. If given a feasible start solution and a test set for a *pure integer (linear) programming* (IP) problem the algorithmic idea is straightforward: iteratively find an element of the test set that maintains feasibility when added to the solution. If no such element exists, the current solution is optimal. When Graver introduced the idea of test sets for integer programs, he showed that finite test sets exist for every feasible IP [83]. Weismantel gives a good overview on different methods to computationally obtain test sets for IPs [172].

The "simplified primal integer programming algorithm" in [175] can be seen as one of the first versions of an *integral basis method* or a *primal cutting plane method* for integer programming. The idea is to solve IPs only by means of the primal simplex algorithm, hence starting from a feasible solution (and an associated basis) and only performing simplex pivots that improve the objective, maintaining primal feasibility and integrality of the basic solution. Since this might not be possible in general, the constraint matrix is manipulated. In [175], this is done by adding a Gomory cut (and its slack variable) for the pivot row when conducting the ratio test of the simplex algorithm. This cut itself will then be chosen as the pivot row instead, and by construction the coefficient of the pivoting variable in the cut and the pivot ratio cancel out. As a consequence, both the cut's slack and the pivot variables take integral values in the new linear system that has been enhanced by one column and one row. A different understanding of this procedure is that it cuts off neighboring fractional points of the incumbent feasible solution until it can make a simplex step that leads to a new incumbent.

Several extensions of this algorithm have been suggested. Notably, Haus et al. present an integral basis method that manipulates the columns of the matrix *without any* cuts being added [90]. By contrast, Letchford and Lodi [109] suggest several enhancements that make Young's algorithm converge more quickly by adding *more* cuts. The main improvements come from separating classes of cuts other than only Gomory cuts and by potentially adding several cuts per round, which is typical in dual cutting plane algorithms.

There is a smooth transition between primal methods and primal heuristics. On the one hand, some primal heuristics such as local branching or proximity search can be modified such that they become complete algorithms; see [62, 64]. In this case, each of their iterations will take an integer solution as an input and have either an improved integer solution or a proof of optimality as output, which is the general concept of primal methods. On the other hand, complete primal methods such as test set algorithms or the integral basis method could of course be run for a limited time as a primal heuristic within a branch-and-bound-based MIP solver.

1.3 Primal Heuristics inside MIP Solvers

In state-of-the-art MIP solvers, primal heuristics play a major role in finding and improving integer feasible solutions at an early stage of the solution process. Knowing good solutions early during optimization helps to prune the search tree and to simplify the problem via dual reductions. Further, knowing a solution proves the feasibility of a problem, and a practitioner might be satisfied with a solution that is proven to be within a certain gap to optimality.

A meaningful experiment in [91] categorized the impact of various components of branch-and-bound-based MIP solvers on the primal and dual sides of the solution process, measured by the primal and dual integral [17], respectively.[3] He considered branching rules, node selection, cutting plane generation, presolving and primal heuristics, and either deactivated all algorithms belonging to a certain component or, in the case of branching rule and node selection, switched to the vanilla default rules of random branching and a depth-first search.

One result of this experiment was that, not surprisingly, primal heuristics have a much larger impact on the primal side of the solution process than on the

---

[3] The primal integral is defined formally in Section 1.7.

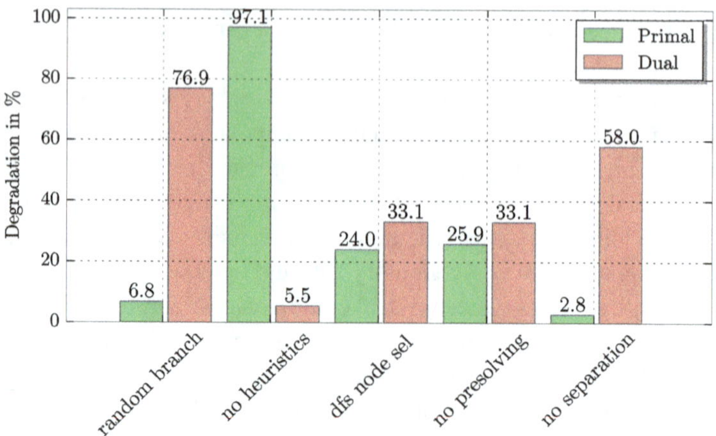

Figure 1.2 Degradation (in percent) of the average primal and dual integral of the MIP solver SCIP when deactivating individual parts of the solver (or changing them to vanilla rules for branching and node selection). Picture by G. Hendel [91].

dual side. Figure 1.2 shows that primal heuristics are by far the most important solver component for achieving good primal performance. Without them, the average primal integral, roughly speaking a measure of the convergence of the primal solution quality, nearly doubles.

These results are in line with those obtained in [6] with a similar experiment, this time based on the commercial MIP solver CPLEX. The results therein show that disabling primal heuristics gives an overall slowdown of approximately 28% in shifted geometric mean, and the slowdown gets higher when restricting to the harder models, that is, models that take more time to solve to proven optimality. Most importantly, primal heuristics play an important role in being able to solve models at all: a non-negligible share of models becomes unsolvable if primal heuristics are disabled.

Heuristics can be called at many different points of the branch-and-bound search. The MIP solver SCIP knows 13 different timings at which external heuristic plugins can be invoked. These include:

- before a node of the branch-and-bound tree is solved;
- after a node of the branch-and-bound tree has been solved;
- after domain propagation, but before LP solving at a node;
- after LP solving, but before branching;
- during LP solving, in each pricing round;
- after the final node of a dive in the branch-and-bound tree;

- various timing points for nodes with LP solves;
- before presolving starts;
- after presolving, before the initial root LP; and
- in between presolving rounds.

Clearly, some overall strategy is needed to coordinate the execution of all the different heuristics within a MIP solver. As usual, a trade-off needs to be sought: on the one hand, we do not want to miss finding good solutions early on in the process; on the other hand, we do not want the primal heuristics to become too time-consuming, in particular on models where they are not very effective.

Common approaches include setting frequencies at which each heuristic should be called, or setting a resource limit on how much should be spent in the solving process executing primal heuristics (e.g., 10% of the running time). In addition, some expensive heuristics are given a chance to run only once during the solution process, usually at the root.

Unfortunately, little is published about these strategies, with two notable exceptions. In [93], multi-armed bandit theory is applied for deciding how to split the heuristic resource budget on the different sub-MIP heuristics implemented in SCIP. The idea is that as we run each of them, we can collect statistics about their success rate on the current model, and bias future calls (within the same solution process) toward the most successful ones, but still giving the other heuristics a chance to run from time to time. Finally, in [27], the overall solution process is formally split into the following three phases:

- feasibility phase, where the solver tries to find a first feasible solution;
- improving phase, where the solver has found a feasible solution, but not yet the optimal one;
- proof phase, where the solver has already found the optimal solution and it *just* needs to prove optimality.

The intuition is that a different tuning of the solver components in general, and of primal heuristics in particular, should be used in the three phases. For example, primal heuristics could be disabled in the last phase, or they could be more aggressive if we are still in the first. Of course, precisely knowing when the transition between the improving and the proof phases happens is a challenge in itself, thus making the strategy adaptation a complex task. The recent line of research on the use of *machine learning* techniques within MIP (see, e.g., [12] for a recent survey on the topic) is a potentially promising direction for leveraging data to tackle the above challenge. Some of the work devoted to the use of modern statistical learning for primal heuristics is discussed in Chapter 8.

Also, it is of utmost importance to balance the time invested versus expected success when designing individual primal heuristics. One common principle of achieving this is the so-called *fast-fail* (or *first-fail*) strategy. This idea has its origins in the constraint programming and artificial intelligence community [89] as a rationale for branching strategies, and was transferred to MIP heuristics in [18]. The rule of [89]: "To succeed, try first where you are most likely to fail" can be interpreted as prioritizing the most critical, or wavering, decisions early on. For variable fixing strategies, for example, this means attempting to fix variables that appear hardest to make feasible first. In a constraint-based view, this means aggressively resolving constraints first that are highly violated, rather than "locking in" constraints that are already feasible or almost feasible. This strategy offers two main benefits. First, it is easier to rectify the implications of a decision early on, when many variables are still unfixed. Second, if those early critical decisions indeed lead to infeasibility, the heuristic fails early, and at least minimizes the wasted computational time, hence the term "fail fast." The opposite strategy – that naively appears more natural to many humans – would be to fix variables or resolve constraints in a way that at each step minimizes the risk of getting an infeasible situation. In many situations, this pushes those hard decisions to the very end, where there is less freedom to resolve their implications, hence making success less likely, and ultimately often leads to running into an infeasible situation only after investing a lot of computational effort.

We end the section by presenting two possible categorizations of primal heuristics. The first categorization is based on the computational expensiveness of the building blocks that the heuristic is allowed to use. In this regard, heuristics inside a MIP solver are usually split into:

- *LP-free heuristics*, that is, those algorithms that are based solely on constraint propagation and other logical reasoning, but that not are allowed to solve linear programs (with the possible exception of solving a final LP in the mixed-integer case after all integer values have been assigned a value).
- *LP-based heuristics*, that is, those algorithms that are allowed to solve LPs, possibly multiple times.
- *sub-MIP heuristics*, that is, those algorithms that even allow the solution of mixed-integer programs, which are typically a (quite) restricted version of the main problem, with strict limits.

Alternatively, we can categorize heuristics depending on their purpose, that is, if their main objective is to construct a feasible solution from scratch (*start* heuristics), or rather to improve upon an already known feasible solution (*improving* heuristics). Typically, but not necessarily, LP-free and and LP-based heuristics are start heuristics, while most sub-MIP heuristics are improving heuristics, defining and exploring neighborhoods associated with the known feasible solutions.

The vast majority of the primal heuristics discussed in the next chapters belong to only one of the classes above, with some notable exceptions. Nevertheless, the heuristics will be presented by grouping them into further (smaller) categories so as to better highlight their characteristics. We will also try to discuss them by referring to the concepts introduced in this section, namely the suitability of each group of heuristics to be invoked in specific points of the branch-and-bound search and its different phases defined above.

1.4 Rounding and Constraint Propagation

The two major components of LP-based heuristics are *rounding* and *constraint propagation*.

1.4.1 Rounding

In the MIP context, we refer to rounding as the operation that is applied to turn an LP-feasible solution $\bar{x}$ into a mixed-integer vector $\hat{x}$. Namely, for each $j \in \mathcal{I}$, if $\bar{x}_j$ is integer, then $\hat{x}_j = \bar{x}_j$. Otherwise, $\hat{x}_j = \lfloor \bar{x}_j \rceil$, which denotes assigning to variable x_j the integer value nearest to the fractional one $\bar{x}_j$. The rest of the solution does not change, that is, for all $j \in \mathcal{N} \setminus \mathcal{I}, \hat{x}_j = \bar{x}_j$. It is easy to see that the resulting vector $\hat{x} \in \mathbb{R}^{n-|\mathcal{I}|} \times \mathbb{Z}^{|\mathcal{I}|}$, but that it does not necessarily satisfy the linear constraints $Ax \leqslant b$ in the MIP (1.1).

The rounding operation is performed on all variables at the same time, and it is the most naive way of trying to recover MIP feasibility from an LP-feasible solution. In practice, rounding is performed sequentially on one or few variables at a time by taking into account its effect, for example, through constraint propagation, the topic of Section 1.4.2.

1.4.2 Constraint Propagation

Constraint propagation is a very general concept that appears under different names in different fields of computer science and mathematical programming. It is essentially a form of inference that consists of explicitly forbidding values – or combinations of values – for some problem variables. Constraint propagation is used as a subroutine inside many practical heuristics, and inside branch and bound as well.

To get a practical constraint propagation system, two questions need to be answered:

- What does it mean to propagate a single constraint? In our particular case, this means understanding how to propagate a general linear constraint with both integer and continuous variables. The logic behind this goes under the name of bound strengthening (a form of preprocessing) in the integer programming community [124, 154].
- How do we coordinate the propagation of the whole set of constraints defining our problem?

In the remaining part of this section we will first describe bound strengthening and then describe the basic concepts of constraint propagation systems, following the propagator-based approach given by Schulte and Stuckey in [158].

Bound Strengthening

Bound strengthening [1, 81, 97, 124, 154] is a preprocessing technique that, given the original domain of a set of variables and a linear constraint on them, tries to infer tighter bounds on the variables. We will now describe the logic behind this technique in the case of a linear inequality of the form

$$\sum_{j \in C^+} a_j x_j + \sum_{j \in C^-} a_j x_j \le b,$$

where C^+ and C^- denote the index set of the variables with positive and negative coefficients, respectively. We will assume that *all* variables are bounded: simple extensions can be made to deal with unbounded (continuous) variables and equality constraints.

For a given linear constraint and a given point $\bar{x}$, we call the value

$$\sum_{j \in C^+} a_j \bar{x}_j + \sum_{j \in C^-} a_j \bar{x}_j,$$

the *activity* of the constraint i with respect to the point $\bar{x}$.

In order to propagate the constraint above, the first step is to compute the minimum ($\underline{\alpha}$) and maximum ($\overline{\alpha}$) *activity level* [41] of the constraint, namely

$$\underline{\alpha} = \sum_{j \in C^+} a_j l_j + \sum_{j \in C^-} a_j u_j,$$

$$\overline{\alpha} = \sum_{j \in C^+} a_j u_j + \sum_{j \in C^-} a_j l_j.$$

Now we can compute updated upper bounds for variables in C^+ as

$$\bar{u}_j = l_j + \frac{b - \underline{\alpha}}{a_j}, \tag{1.3}$$

and updated lower bounds for variables in C^- as

$$\bar{l}_j = u_j + \frac{b - \underline{\alpha}}{a_j}. \tag{1.4}$$

Moreover, for variables constrained to be integer, we can also apply the floor $\lfloor \cdot \rfloor$ and ceiling $\lceil \cdot \rceil$ operators to the new upper and lower bounds, respectively. The maximum activity level is used analogously for constraints in $\geq$ form.

Let's consider, for example, the following system of linear constraints over five variables:

$$\begin{aligned}
3x_1 + 2x_3 + x_4 &\leq 5, \\
x_4 - 6x_5 &\geq 0, \\
x_1 + x_2 - x_5 &\leq 0, \\
x_1 &\in \{0, 1\}, \\
x_2 &\in \{0, 1\}, \\
x_3 &\in \{0, \ldots, 5\}, \\
x_4 &\in \{0, \ldots, 10\}, \\
x_5 &\in \{0, 1\}.
\end{aligned} \tag{1.5}$$

The minimum activity $\underline{\alpha}$ of the first constraint is zero, and that gives updated upper bounds $x_4 \leq 5$ and $x_5 \leq 2$ (for the last upper bound, integrality is exploited). Then, the updated maximum activity $\bar{\alpha}$ for the second constraint implies the upper bound $x_5 \leq 0$ (again, exploiting integrality). Finally, the last constraint derives $x_1 = x_2 = 0$. Note that bound strengthening, being allowed to exploit integrality information, can derive bounds that are invalid for the linear programming relaxation, and thus strengthen the model: in the example above, the fractional solution $(0, 5/6, 0, 5, 5/6)$ is violated by the derived bounds.

It is worth noting that no propagation is possible in the case that the maximum potential activity change due to a single variable, computed as

$$\max_j \{|a_j(u_j - l_j)|\},$$

is not greater than the quantity $b - \underline{\alpha}$. This observation is very important for the efficiency of the propagation algorithm, since it can save several useless propagator calls.

Finally, for linear constraints of special form, there often exist stronger or faster propagation algorithms. For example, a knapsack constraint, where all variables are binary and all coefficients are integer, can be propagated by making use of integer arithmetic instead of floating point arithmetic. Set covering constraints, that is, constraints of the form

$$x_{j_1} + x_{j_2} + \cdots + x_{j_k} \geq 1,$$

where all variables are binary, admit only one kind of propagation, namely fixing the remaining variable to 1 once all the others appearing in the constraint are fixed to 0, and propagating a set of set-covering constraints can be implemented very efficiently via the so-called *two-watched literals scheme* [133]. For an overview of specialized propagators for linear constraints, we refer to [1].

Propagation Algorithm

Now that we have described what it means to propagate a single linear constraint, we move on to describe how the propagation of the different constraints is coordinated within a MIP solver. The basic idea is that one does not just need to propagate all linear constraints once, as reductions obtained when propagating one constraint can lead to further reductions from constraints that we already propagated. Intuitively, we need to keep propagating the relevant constraints until a fix point is reached, that is, no more propagations can be found. It is also important to note that, in general, we do not just use the linear constraints of the model during propagation, but also other related global structures, for example, the clique table or the implication graph (see, e.g., [1] and Section 2.5).

A general framework for describing the behavior of (efficient) constraint propagation systems can be found in the constraint programming [149] literature, most notably in [157, 158]. In the remaining part of this section, we will introduce the general concepts and describe how they map to the MIP case.

Constraint propagation systems are built upon the basic concepts of *domain*, *constraint* and *propagator*.

A *domain* D is the set of values a solution x can possibly take. In our case, the domain is defined by

$$
\begin{aligned}
l_j \leqslant x_j \leqslant u_j, & \quad \text{for all } j \in \mathcal{N}, \\
x_j \in \mathbb{R}, & \quad \text{for all } j \in \mathcal{N} \setminus \mathcal{I}, \\
x_j \in \mathbb{Z}, & \quad \text{for all } j \in \mathcal{I}.
\end{aligned}
\tag{1.6}
$$

In full generality, a *constraint* c is a relation among a subset of variables listing the tuples allowed by the constraint itself. However, this (very general) definition is of little use from the computational point of view. In the MIP case, constraints are obviously defined as the linear constraints of the model, but in general we also propagate other structures: those also play the role of constraints during propagation. In order to have a more practical, yet still general, abstraction, constraint propagation systems implement constraints through so-called *propagators*.

A propagator p implementing a constraint c is a function that maps domains to domains and that satisfies the following conditions:[4]

- p is a *decreasing* function, that is, $p(D) \subseteq D$ for all domains. This guarantees that propagators only remove values.
- p is a *monotonic* function, that is, if $D_1 \subseteq D_2$, then $p(D_1) \subseteq p(D_2)$.
- p is *correct* for c, that is, it does not remove any tuple allowed by c.
- p is *checking* for c, that is, all domains D corresponding to solutions of c are *fixpoints* for p, that is, $p(D) = D$. In other words, for every domain D in which all variables involved in the constraint are fixed and the corresponding tuple is valid for c, we must have $p(D) = D$.

Again, propagators can be associated not just with the linear constraints of the model, but with other global structures such as the clique table or the implication graph. In addition, propagators can be used to implement *dual* reductions, that is, bound changes that are not necessarily valid for all feasible solutions of the model at hand, but that still guarantee that at least one optimal solution is left. The most common example is the so-called reduced cost fixing [140], but others such as *orbital fixing* [144] are also common in modern MIP solvers.

A *propagation solver* for a set of propagators R and some initial domain D finds a fixpoint for propagators $p \in R$.

A basic propagation algorithm is outlined in Algorithm 2. On input, the propagator set is partitioned into the sets P_f and P_n, depending on the known fixpoint status of the propagators for domain D – this feature is essential for implementing efficient incremental propagation. The algorithm maintains a queue Q of pending propagators (initially P_n). At each iteration, a propagator p is popped from the queue and executed. At the same time, the set K of variables whose domains have been modified is computed and all propagators that share variables with K are added to Q (hence they are *scheduled* for execution).

The complexity of this algorithm is highly dependent on the domain of the variables. For integer (finite domain) variables, the algorithm terminates in a finite number of steps, although the complexity is exponential in the size of

---

[4] In general, a constraint c is implemented by a collection of propagators; we will consider only the case where a single propagator suffices.

---

input : a domain D
input : a set P_f of (fixpoint) propagators p with $p(D) = D$
input : a set P_n of (non-fixpoint) propagators p with $p(D) \subseteq D$
output: an updated domain D

1 $Q = P_n$
2 $R = P_f \cup P_n$
3 **while** Q *not empty* **do**
4      $p = \text{Pop}\,(Q)$
5      $D = p(D)$
6      $K = $ set of variables whose domain was changed by p
7      $Q = Q \cup \{q \in R : var(q) \cap K \neq \emptyset\}$
8 **end**
9 **return** D

Algorithm 2   Basic propagation engine.

the domain (it is however polynomial in the pure binary case, provided that the propagators are also polynomial, which is usually the case). For continuous variables, this algorithm may not converge in a finite number of steps, as shown in the following example (Hooker [97]):

$$\begin{cases} \alpha x_1 - x_2 \geq 0, \\ -x_1 + x_2 \geq 0, \end{cases} \tag{1.7}$$

where $0 < \alpha < 1$ and the initial domain is $[0, 1]$ for both variables; it can be easily seen that the upper bound on x_1 converges only asymptotically to zero. In a first round of propagation, the upper bound of x_2 would be tightened to α due to the first inequality, and in turn the upper bound of x_1 would be tightened to α due to the second inequality. In the next round, the upper bound of x_2 would be tightened to α^2 due to the first inequality, leading to an upper bound of α^2 for x_1, and so on, as shown in Figure 1.3. So, in practice, Algorithm 2 is stopped after some predefined number of iterations or if the reduction in the domain of variable falls below some given threshold.

1.5 General Heuristic Concepts

While primal heuristics for MIP clearly exploit the specific properties and tools of the MIP paradigm, for example, the global view of the problem given by the MIP formulation and the ability to solve LP relaxations or even related

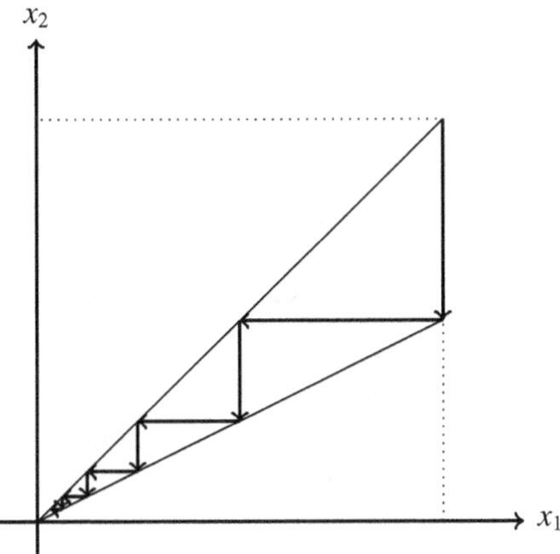

Figure 1.3 A nasty example for constraint propagation.

(sub-)MIPs, they can often be described and interpreted with the language of general mathematical heuristics for combinatorial optimization problems. In particular, many constructive heuristics follow (more or less strictly) a greedy approach, while most improvement heuristics are based on local search. In this section, we give a general overview and definition of those basic concepts.

1.5.1 The Greedy Paradigm

A very natural approach for solving combinatorial problems is to use a strategy that takes a sequence of decisions, with a set of different possible choices at each step. If at each step we always pick the choice that is locally optimal, that is, it is the best choice given the decisions we have made so far, and we never backtrack on those decisions, then we have a *greedy* algorithm [48, chapter 16].

Greedy algorithms are not guaranteed to return the optimal solution (or even a feasible solution) to an optimization problem, but in some cases they do. The effectiveness of a greedy approach by far depends on the structure of the problem: for some specific classes of problems, such as minimum spanning tree, shortest path and continuous knapsack (to name a few), there are greedy algorithms that solve the problem to proven optimality. In less ideal, but still good cases, they always return a feasible solution and are able to guarantee a constant worst-case approximation ratio; for example, the greedy algorithm returns a solution that is at most $1 - 1/e$ away from the optimum in submodular function maximization [141], where e is the Euler constant. Similarly, with some care, the greedy strategy gives a 2-approximation for binary knapsacks [125]. For other combinatorial problems, the best guarantee is no longer constant: the approximation is $O(\log n)$ for both the set covering problem and the *traveling salesman problem* (TSP), where the heuristic is known as nearest neighbor. Finally, no guarantee whatsoever can be given for a general MIP. Another substantial benefit of greedy strategies, besides their simplicity, is that they are typically quite fast. With appropriate data structures, greedy algorithms can be often implemented in linear time, potentially with an additional logarithmic factor in case sorting is needed. So, even if they are not guaranteed to find a solution, they are usually worth a try.

Inside MIP solvers, diving heuristics, that is, methods that simulate a dive in the branch-and-bound tree by iteratively fixing a variable and resolving the LP relaxation, are prime examples of greedy strategies.

1.5.2 Local Search

Another approach for designing heuristics for a given optimization problem is that of local search. Again, the concept is very natural: given a known feasible solution, often referred to as "reference" or "candidate" solution, it is usually worth looking for a better one in a properly defined neighborhood in the solution space. Then, the approach can be repeated until a locally optimal solution is found.

For structured optimization problems, local search heuristics rely on ad hoc definitions of neighborhood, based on so-called *moves*. For example, permutation problems (e.g., the TSP) are amenable to 2-opt neighborhoods, where a solution is in the neighborhood of the reference solution if it can be obtained from the latter by swapping two elements in the sequence. Another (even simpler) neighborhood is 1-opt: given a reference solution, this is defined as the set of solutions that can be obtained by changing the value of a single variable. Note that a 1-opt neighborhood might not contain any additional solution (e.g., in the TSP case). On the one hand, the size (and structure) of the neighborhood has a direct effect on the complexity of the method. Small neighborhoods are clearly faster to explore, but the chances of finding an improving solution are smaller, so there is a higher risk of getting stuck. On the other hand, very large neighborhoods can be prohibitively expensive to explore, at least exhaustively.

In the MIP case, we often try to sidestep the issue by applying a so-called large neighborhood search (LNS) approach, in which we define a neighborhood implicitly by adding constraints (often, but not necessarily, variable fixings), and then apply MIP technology recursively on the resulting sub-MIP. While this is quite elegant, it is still quite challenging to predict a priori how expensive each such sub-MIP is to solve, so proper limits and an outer logic to adjust the size of the subproblems is needed to get a practical method (see Chapter 2).

Independently from the choice of neighborhood, we also need a sound strategy to escape local optima, and we have a great variety of options, namely:

- we can restart the local search from a different solution, obtained by sampling or perturbation;
- we can change the acceptance criterion when exploring a neighborhood, allowing for non-improving solutions to be accepted;
- we can change the neighborhood (either in size or structure);
- all of the above.

There is a vast literature in the *meta*heuristic community [78] on strategies based on local search: iterated local search (ILS), tabu search, variable neighborhood search (VNS), simulated annealing and random walks, just to name a few, have all proven quite successful on specific optimization problems, and are tools of wide applicability, although they often require extensive tuning to give the best results on a given class of problems. We note though that those meta approaches, while crucial when designing standalone heuristics from scratch, are less important within the MIP framework, where we rely on branch and bound itself to let the heuristics escape local optima.

Finally, note that local search is in many ways complementary to greedy methods: greedy strategies can be used to find feasible solutions, while a subsequent local search phase (i) improves their quality, and (ii) makes sure that those are at least locally optimal.

1.5.3 Population-Based Methods

The local search methods described in Section 1.5.2 are so-called single solution methods: they move from feasible solution to feasible solution, and the currently explored neighborhood is defined starting from the current solution only. These methods are of course allowed to (and often do) store past solutions in a so-called *solution pool*, but those do not affect the behavior of the algorithm. By contrast, population-based methods maintain a set of feasible solutions at each iteration, and they explore the solution space based on some properties of the whole set.

The most well-known population-based algorithms by far are *genetic algorithms* [132] that are inspired by natural evolution: at each iteration, the current generation of individuals (i.e., set of solutions) is mutated/combined to obtain the next generation, and selection is applied according to fitness (i.e., the objective function). While the intuition of evolutionary algorithms is clear, their design is less trivial: many different steps need to be specified, often in a problem-specific fashion, and there are many parameters to be tuned in order to get the best results on a given class of problems. However, once designed, they are also, in general, quite easy to implement and they parallelize trivially.

Evolutionary algorithms are not predominant in MIP, for the same reason as for local-search meta schemes, but some limited form of them, based on the MIP solution pool, is actually implemented in all state-of-the-art MIP solvers.

1.5.4 Diversification vs. Intensification

To be successful, a primal heuristic needs to be able to find good quality solutions in an exponentially sized solution space, and to do so in a reasonable amount of time. In order to achieve such goal, when designing a primal heuristics two opposite goals must be continuously kept in balance. On the one hand, we want to sample different parts of the solution space in order to reduce the chances of getting stuck in an unpromising region: this goal is called diversification, and it is often achieved through a random component. On the other hand, we also need to invest more resources in the promising parts, as sampling alone is not going to find the proverbial needle in a haystack quickly enough on average; this is called intensification, and it is often (although not necessarily) based on some form of local search.

Most of the heuristics that we will describe in the book can be analyzed under the lens of diversification vs. intensification. A greedy algorithm, for example, does neither; it neither diversifies (to the contrary, being deterministic it would always yield the same outcome), nor intensifies, and this is why a single greedy approach is rarely a good option on general problems. At least on the first front, the greedy scheme can be improved by incorporating randomization. The most common approach is the so-called GRASP scheme [59], where at each step, instead of just picking the locally optimal solution, we build a shortlist of best candidates and choose randomly within them. To the contrary, a pure local search method intensifies very well but lacks diversification; indeed, all the meta schemes mentioned in the previous sections can be interpreted as a way to add diversification to the basic method (and thus escape local optima).

1.6 Reference Points, Information and Statistics

In this section, we introduce some extra definitions and concepts that help discuss and analyze primal heuristics for MIP.

1.6.1 Reference Points

A variety of reference points are of interest for primal heuristics.

The *incumbent solution*, that is, the best feasible solution that has been found so far during a running MIP solve, is used as a reference point for the vast majority of improvement heuristics. In particular, large neighborhood search heuristics such as RINS [51], proximity search [65], or local branching [62] make use of the incumbent. Guided diving [51] is an example of a diving heuristic that requires an incumbent solution.

Other feasible solutions are of comparable interest. Crossover is a large neighborhood search heuristic that requires more than one solution. Furthermore heuristics might use infeasible integer points as reference, for example, feasibility pump or some variants of local branching or proximity search. Finally, a partial assignment of the integer variables might be used to initialize local search heuristics. For example, this can be provided as an input to a MIP solver.

Just as every MIP might have many different optimal solutions, there might also be multiple optima of the LP relaxation. Many MIP heuristics consider an optimal solution of the LP relaxation as a reference point. Besides that, alternative optimal or feasible solutions for the LP relaxation might be used to guide heuristics. Another important reference point is the so-called analytic center.

The analytic center x^{ac} of a bounded polyhedron given in equality form ($Ax = b, x \geqslant 0$) has been introduced by Sonnevend [163] and is defined as

$$x^{\text{ac}} = \operatorname{argmin} \left\{ -\sum_{j \in N} \ln x_j : Ax = b \right\}. \tag{1.8}$$

The analytic center can be efficiently computed by using a barrier algorithm. Note that the strong convexity of the logarithm implies that the analytic center of a bounded polyhedron is uniquely defined. It maximizes the distance to the boundary due to the logarithm tending to minus infinity when its argument is going to zero. If the polyhedron is a simplex, the analytic center is also the barycenter of the polyhedron [164].

1.6.2 Solving Statistics

The *pseudocosts* [13, 72] of a variable are a statistic that is collected during the course of a branch-and-bound algorithm. Pseudocosts track the history of how much the dual bound improved when branching on a given variable in previous nodes. These statistics can then be used to estimate how much the objective will change when the bounds of the variables are tightened.

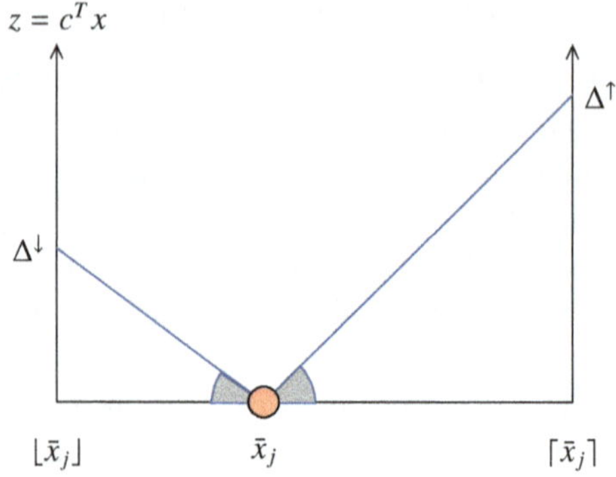

Figure 1.4 Graphical representation of pseudocosts update.

Pseudocosts are updated after a bound change of a variable has been performed (e.g., due to branching or a fix in a diving heuristic) and the LP relaxation of the tightened problem has been solved. When updating the pseudocosts, the objective gain per unit change in variable x_j is computed. Assume we change the lower bound of x_j, that is, we rounded it up. Then, the upward gain ς_j^+ is computed as

$$\varsigma_j^+ := \frac{\Delta^{\uparrow}}{\lceil \bar{x}_j \rceil - \bar{x}_j}, \tag{1.9}$$

where $\Delta^{\uparrow}$ is the difference between the optimal LP objectives of the subproblem with x_j rounded up and the optimal LP objective before conducting the rounding. The downward gain ς_j^- when rounding down a variable is computed accordingly. The thin, light blue line in Figure 1.4 illustrates the operation. These estimation formulas are based on the assumption that the objective increases linearly in both directions.

Let σ_j^+ denote the sum of ς_j^+ over all subproblems that were created by branching upward on x_j and whose LP relaxation has already been solved and was feasible. Further, let v_j^+ be the number of such problems. Then the upward pseudocosts of variable x_j are calculated as the arithmetic mean of all objective gains (per unit step length) observed by branching upward on x_j, namely

$$\Psi_j^+ := \frac{\sigma_j^+}{v_j^+}. \tag{1.10}$$

The downward pseudocosts Ψ_j^- are computed analogously from subproblems that were created by branching downward on x_j and whose LP relaxation has already been feasibly solved.

The *upward inference value* of a variable x_j, $j \in \mathcal{I}$, is defined analogously to the pseudocosts as

$$\tilde{\varsigma}_j^+ := \frac{\tilde{\sigma}_j^+}{\tilde{\nu}_j^+}, \tag{1.11}$$

where $\tilde{\sigma}_j^+$ is the sum of the number of domain reductions found by propagation taken over all subproblems that were created by branching upward on x_j for which domain propagation has already been applied, and $\tilde{\nu}_j^+$ is the number of such subproblems. Again, the downward inference $\tilde{\varsigma}_j^-$ value is defined analogously.

Branching statistics are useful information for primal heuristics to make decisions, but heuristics can also contribute to these statistics. Rapid learning [21] takes this to the extreme. It is applies a fast depth-first branch-and-bound search without solving LPs for a limited number of nodes, initializing, among other statistics, inference values.

1.7 Measuring the Impact of Primal Heuristics

When implementing optimization software, two questions naturally arise: how does the new code perform with respect to existing code, and which are the best settings for a particular algorithm? This goes back to the early days of operations research. Hoffman et al. reported a first computational experiment to compare different implementations of linear programming algorithms in 1953 [95]. Just as researchers and software vendors want to distinguish their code on general test sets, a user wants to tune their optimization software for a particular set of problems. However, all parties require suitable criteria for measuring the performance of a software implementation.

In mathematical programming, the running time to optimality, the number of branch-and-bound nodes and the number of simplex or interior point iterations are commonly used performance measures. All of these mostly depend on the convergence of the dual bound since, in practice, it typically takes much longer until the dual bound converges to the optimal value than it takes to find a primal solution of the optimal value. This is not only an empirical observation, but also indicated by complexity theory. Finding a solution for a certain objective value is $\mathcal{NP}$-complete, but proving a certain bound is co-$\mathcal{NP}$-complete and thereby most likely not even in $\mathcal{NP}$, is hence arguably harder! Feasibility certificates (hence solutions) are of linear size, but a MIP optimality certificate is not – it can be an exponential tree or set of cuts.

Furthermore, measures such as time to optimality or number of branch-and-bound nodes make most sense for instances that can actually be solved within a given time limit – whereas employing heuristics is particularly worthwhile for hard instances that cannot be solved to proven optimality within a reasonable time.

For primal heuristics, it seems logical to consider measures that consider primarily the primal bound and that are independent of an eventual timeout. Examples of such measures are the time needed to find a first feasible solution, an optimal solution, or a solution within a certain gap to optimality (see, e.g., [96]). Each of these has its individual strengths and weaknesses. The time to first solution entirely disregards the solution quality: for about one quarter ($^{23}/_{87}$) of the MIPLIB 2010 [106] benchmark instances, a trivial solution of all variables set to their lower bound (or all to their upper bound) is feasible, but most of the time such a solution does not provide valuable information to the user – it corresponds to obvious extreme cases such as utilizing all available resources when trying to minimize the number of used resources or producing no goods when trying to maximize the amount of produced goods. Particularly when analyzing heuristics embedded in a complete solver, the time to the first solution mainly measures the time needed for preprocessing and solving the root node relaxation; the MIP solvers CPLEX, GUROBI and XPRESS find solutions for the vast majority of the MIPLIB 2010 benchmark instances during root node processing. Instead, the time to optimal solution ignores that slightly suboptimal but practically sufficient solutions might have been found long before. Finally, taking the time to a certain gap is an attempt to balance this, but the choice of the threshold is arbitrary by design.

Altogether, the most important consideration for primal heuristics is the trade-off between speed and solution quality. None of the above performance measures gives a complete assessment of this trade-off. In [17], Berthold introduces a new performance measure that takes into account the whole solution process, and is especially targeted at benchmarking primal heuristics. It measures the progress of the primal bound's convergence toward the optimal solution over the entire solving time.

Assume the objective function values of intermediate incumbent solutions and the points in time when they have been found to be given – for a particular MIP solver, a certain problem instance and a fixed computational environment. This information can be gathered from the log files that standard MIP solvers produce.

Definition 1.6 (primal gap function)　Let $t_{max} \in \mathbb{R}_{\geq 0}$ be a limit on the solution time of a MIP solver. Its *primal gap function* $p \colon [0, t_{max}] \mapsto [0, 1]$ is defined as follows:

$$p(t) := \begin{cases} 1, & \text{if no incumbent until time } t, \\ \gamma^p(\tilde{x}(t)), & \text{with } \tilde{x}(t) \text{ being the incumbent at time } t, \text{ otherwise.} \end{cases}$$

Here, γ^p is the primal gap as defined in Definition 1.3. The primal gap function $p(t)$ is a step function that changes whenever a new incumbent is found. It is monotonically decreasing, one at $t = 0$, and zero from the point at which the optimal solution is found.

Definition 1.7 (primal integral)　Let $T \in [0, t_{max}]$ and let $t_i \in [0, T]$ for $i \in \{1, \dots, I - 1\}$ be the points in time when a new incumbent solution is found, $t_0 = 0$, $t_I = T$. The *primal integral* $P(T)$ of a run is defined as

$$P(T) := \int_{t=0}^{T} p(t)\, dt = \sum_{i=1}^{I} p(t_{i-1}) \cdot (t_i - t_{i-1}).$$

The fraction $P(t_{max})/t_{max}$ can be seen as the average solution quality during the search process. In other words, the smaller $P(t_{max})$ is, the better is the expected quality of the incumbent solution if the solver is stopped at an arbitrary point in time.

Berthold [17] suggests using $P(t_{max})/t_{max}$ for measuring the quality of primal heuristics. This measure features two simple, but important attributes. First, whenever a better solution is found at the same point in time, $P(t_{max})$ decreases. Second, whenever the same solution is found at an earlier point in time, $P(t_{max})$ decreases. Briefly, the primal integral favors finding good solutions early. For the performance measures discussed so far, at most one of these two attributes holds in general.

Consider Figure 1.5 for a visualization of the primal integral. The thick red line shows the development of the primal bound when using heuristics; the red-shaded area corresponds to the primal integral of that solution process. The thick green line shows the development of the primal bound when not using heuristics; the green-shaded (plus the red-shaded) area corresponds to the primal integral of that solution process. The primal integral of the run with heuristics is smaller (hence better) since solutions with a small gap are found earlier.

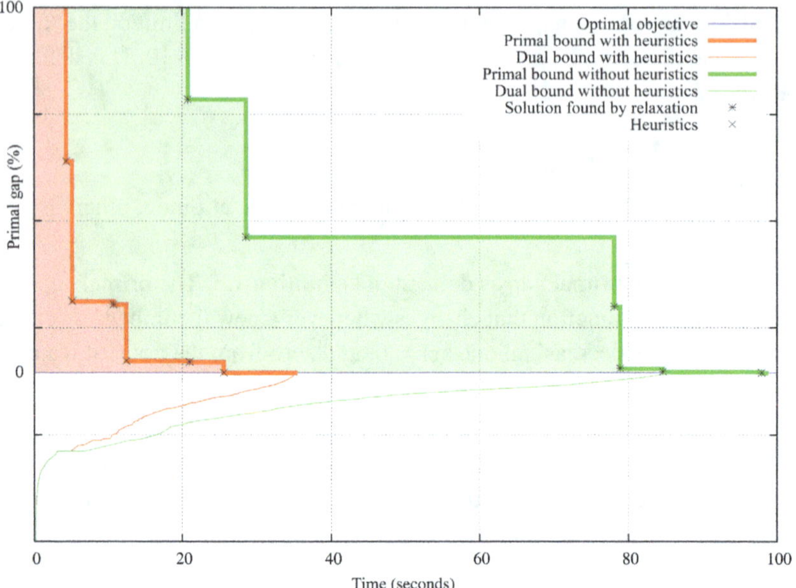

Figure 1.5 Visualization of the primal integral for two solution processes of the same instance.

1.8 Quiz

It's quiz time! This section is designed to test your understanding of the concepts covered in this chapter. Each question is followed by four possible answers, only one of which is correct. If you are unsure about some questions, you might find it helpful to work through the corresponding section again. Remember, sometimes the process of elimination can be as valuable a tool as direct problem-solving. By engaging with this quiz, you'll reinforce your knowledge and identify areas requiring further study. Take your time, think carefully and check the answers in the Appendix when you are done.

Question 1 The feasible region of the LP relaxation of a MIP is a polyhedron

☐ only if all variables are bounded;
☐ always;
☐ only if there are no equalities;
☐ never.

Question 2 Consider a simple binary knapsack problem in standard form (all positive data):

☐ all variables are up-locked;

☐ all variables are down-locked;

☐ all variables have zero locks;

☐ all variables are both up- and down-locked.

Question 3 In an LP-based branch-and-bound algorithm, the primal-dual gap between the current best primal bound and current best dual bound

☐ changes erratically;

☐ is monotonically decreasing and reaches zero when the algorithm terminates;

☐ is monotonically increasing and reaches one when the algorithm terminates;

☐ stays constant, as it is a property of the instance;

Question 4 Within the context of branch and bound, primal heuristics are

☐ useless, branch and bound is an exact method and does not need heuristics;

☐ required for branch and bound to converge;

☐ practically relevant and effective at reducing the primal gap;

☐ irrelevant for the optimality proof.

Question 5 Consider an integer program with all binary variables. Constraint propagation is

☐ a complete and polynomial inference scheme;

☐ a complete but non-polynomial inference scheme;

☐ an incomplete but polynomial inference scheme;

☐ an incomplete and non-polynomial inference scheme.

Question 6 In a MIP context, a greedy heuristic followed by a local search procedure

☐ is guaranteed to find a feasible solution;

☐ always finds the optimal solution if the local search neighborhood is a 2-opt;

☐ is guaranteed to find a feasible solution with a $O(\log n)$ worst-case approximation ratio;

☐ has no guarantees.

Question 7 The primal integral

☐ is independent of the chosen time limit;

☐ can only be approximated with a finite summation;

☐ is a measure balancing solution quality and runtime;

☐ is always a number in $[0, 1]$.

2

Large Neighborhood Search

The vast majority of the primal heuristics that will be described in this chapter have two main characteristics in common.

First, they assume the knowledge of one or more feasible solutions so as to define and explore a suitably crafted neighborhood. In that sense, they are local search heuristics, a category informally introduced in Section 1.3 and whose basic ingredients were discussed in Section 1.5.2. Indeed, this feature highlights the connection between the class of primal heuristics presented here and the vast literature on local search, which is not limited to the context of MIP solvers. More precisely, local search primal heuristics have always been a basic component of MIP solvers because they borrow seminal ideas from the classical 1-opt and 2-opt neighborhoods (see Section 1.5.2) to compute approximate solutions for most problems in combinatorial optimization, the traveling salesman problem being the most famous one.

Second, the neighborhoods explored by the primal heuristics considered in this chapter are generally "large," that is, they cannot be explored by complete enumeration without incurring very long computing times, which is clearly incompatible with the idea of invoking those algorithms within the branch-and-bound search (recall the discussion in Section 1.3). Again, this is a challenge that the local search literature encountered: large neighborhoods have higher chances of containing good, improving solutions, at the cost of requiring large computing times to explore them. Within these lines, many examples of large neighborhoods with problem-dependent characteristics, so that they can be explored in reasonable computing times, have been presented in the literature, including the very elegant area of the *very large neighborhood search* [7]. However, problem-dependent ideas are rarely compatible with the MIP setting and, in this particular case, somehow innovation flowed in the opposite direction. Namely, it is in the MIP context that the ideas of (i) defining (often through linear inequalities) general large neighborhoods and (ii) exploring

30

them by recursively calling a MIP solver [62] originated, and afterwards were borrowed by the local search community, to define the entire new area of the so-called *Mat*heuristics [123].

All of the primal heuristics presented in this chapter explore large neighborhoods by means of calls to a "black-box" MIP solver, a strategy that turns out to be effective and efficient, through artificially limiting its computational effort, in improving a reference solution (generally, the incumbent) or a population of solutions. They differ in the way the problem-*in*dependent neighborhoods – treated as general MIP instances – are defined. These differences need to take into account the trade-off between the richness of the neighborhood, that is, the chance it contains improving solutions, and its complexity, that is, the computational effort spent in exploring it.

Using the local search terminology introduced in Sections 1.5.2 and 1.5.4, we will present the *large neighborhood search* (LNS) [161] heuristics in this chapter by highlighting for each of them the design choices for the three classical local search building blocks, namely (i) *neighborhood* definition, (ii) *intensification* approach, and (iii) *diversification* strategy. The LNS algorithms presented in this chapter are MIP primal heuristics because:

i) the neighborhood is constructed through MIP modeling tools (linear inequalities, bound changes, etc.); that is, it defines a new MIP instance (generally referred to as a sub-MIP);

ii) the intensification corresponds to exploring the neighborhood by invoking a MIP solver (generally by using appropriate work limits, the most common being a limit in the number of branch-and-bound nodes);

iii) the diversification is implemented by integrating the LNS within the MIP branch-and-bound scheme.

2.1 Local Branching

In the *local branching* framework [62], originally introduced for MIP problems in which at least a group of variables $\mathcal{B} \subseteq I$ is binary, a neighborhood was defined with respect to a reference (feasible) solution $\tilde{x}$ and a positive integer k by the Hamming distance constraint

$$\Delta(x, \tilde{x}) = \sum_{j \in \mathcal{B}: \tilde{x}_j = 0} x_j + \sum_{j \in \mathcal{B}: \tilde{x}_j = 1} (1 - x_j) \leqslant k. \tag{2.1}$$

The linear constraint (2.1) defines the neighborhood of all feasible solutions whose Hamming distance in the binary support of $\tilde{x}$ is at most k. Thus, the

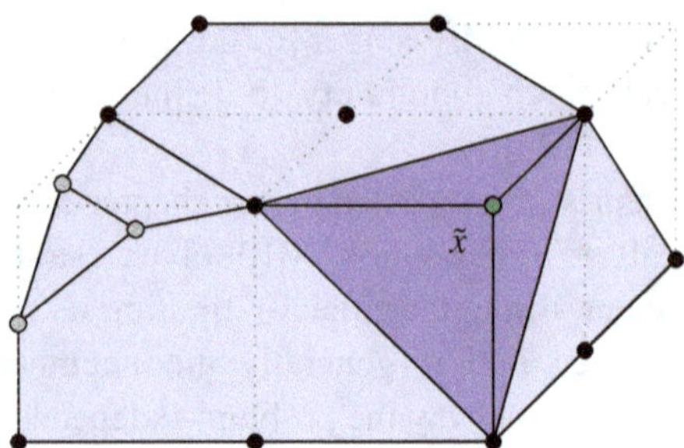

Figure 2.1 Local branching sub-MIP on a small three-dimensional example. Fractional vertices are marked in gray, while integer vertices are marked in black.

local branching constraint (2.1) is clearly invalid for the original MIP, but it is used to define a sub-MIP that is then solved by calling the same MIP solver. A pictorial representation of the local branching sub-MIP on a small three-dimensional example is given in Figure 2.1.

It is easy to see that the characteristics of the sub-MIP both in terms of the richness of improving solutions and in terms of practical computational complexity depend on the value of k, where obviously the larger the value of k, the higher the chances of finding improving solutions and the more time it takes to solve the sub-MIP. However, the intuition behind local branching is that the search for improving solutions around the reference one is both well suited for MIP technology and facilitated by the introduction of constraint (2.1). Indeed, it is easy to see that for $k = |\mathcal{B}|$ the local branching constraint is redundant in the sub-MIP, while, for more restrictive k values,[1] it immediately becomes tight in the root LP relaxation (likely lifting the lower bound with respect to that of the original MIP), and for the MIP solver it is natural to enumerate the binary variables that are fractional in that LP solution. Back in the early 2000s, the "gamble" relied on considering MIP technology to be mature enough to make such an implicit enumeration effective with respect to a classical full enumeration. In practice, MIP solvers are indeed effective in solving sub-MIPs associated with large neighborhoods, and the improving solutions, if they exist, are generally found early in the search tree (see the beginning of this chapter for the work limit of the sub-MIP solution), though proving optimality might remain challenging.

In the original version of [62], the local branching framework was presented in a general form that allowed its use outside of a MIP solver context as a

---

[1] The choice of setting for all problems in the benchmark, $k = 20$ in [62], was effective, though purposely crude. In [113], such a choice is revisited through a machine learning lens; see Section 8.2.1.

standalone exact and/or heuristic algorithm. This way of presenting it is outside the scope of this book and the interested reader is referred to [62] for the details. However, it is as a primal heuristic that the local branching idea has caught the attention of the MIP community, and it is currently implemented in different ways in most of the solvers. Notably, these implementations rely on the idea of invoking local branching when a new incumbent solution is obtained (either by another primal heuristic or simply by branching) to locally improve it. The associated sub-MIP is solved for a prefixed number of branch-and-bound nodes (often the magic number 500), so as to limit its computational footprint. Running the MIP solver that explores the neighborhood with a node limit is common practice for all the methods we will discuss in this chapter.

Finally, constraint (2.1) can be generalized to the case of general integer variables [62]. Namely,

$$\sum_{j \in I:\tilde{x}_j=l_j} (x_j - l_j) + \sum_{j \in I:\tilde{x}_j=u_j} (u_j - x_j) + \sum_{j \in I:l_j<\tilde{x}_j<u_j} (x_j^+ + x_j^-) \leqslant k, \qquad (2.2)$$

where $x_j = \tilde{x}_j + x_j^+ - x_j^-$, $x_j^+ \in [0, u_j - \tilde{x}_j]$, $x_j^- \in [0, \tilde{x}_j - l_j]$ and $x_j^+, x_j^- \in \mathbb{Z}$. Essentially, the generalization[2] requires the introduction of artificial variables to deal with those variables in $\tilde{x}$ that are not at their bounds and highlights the reason why binary variables are cleaner to manage in MIP. Admittedly, local branching has proven to be especially powerful for MIP problems in which a subset of binary variables exists, and deciding on such a subset is of fundamental importance for producing good solutions. In fact, this requirement is not very restrictive in practice; this is true in many applications, for example, all variants of *facility location* problems.

2.2 Relaxation Induced Neighborhood Search

The paper that introduced the *relaxation induced neighborhood search* (RINS) heuristic [51] made a significant step forward with respect to the integration of LNS heuristics within MIP by explicitly formalizing the abstract concepts discussed at the end of the introduction to this chapter; that is, neighborhood definition (through MIP tools), intensification by MIP solving and diversification within the branch-and-bound tree. In particular, RINS defines the neighborhood by comparing an initial feasible solution $\tilde{x}$ with a solution of the LP relaxation $\bar{x}$ and fixing any variable x_j, $j \in I$, such that $\tilde{x}_j = \bar{x}_j$. The neighborhood, in the form of a sub-MIP, is explored by simply calling a MIP solver in a black-box manner with work limits.

---

[2] Each term of constraint (2.2) can be normalized in the attempt to make the left-hand side comparable with k. The reader is referred to [62] for a more complete description.

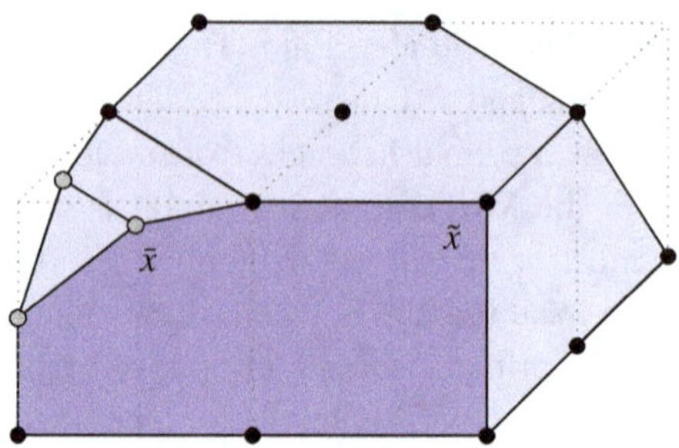

Figure 2.2 RINS sub-MIP on a small three-dimensional example. Fractional vertices are marked in gray, while integer vertices are marked in black. The $\tilde{x}$ solution is the same as in Figure 2.1.

A pictorial representation of the RINS branching sub-MIP on the same small three-dimensional example is given in Figure 2.2. Compared with local branching, hard-fixing variables[3] reduces the size of the sub-MIP, with a potential loss of flexibility, resulting in problems that are less computationally intense. Remarkably, the use of the solution of the LP relaxation $\bar{x}$ provides a natural diversification mechanism because $\bar{x}$ changes at every node of the branch-and-bound tree, so, potentially, RINS can be invoked frequently. Another characteristic of the method is that, in its simplest form, it does not depend on any parameter (such as, for example, the value k in a local branching constraint): on the one hand, this makes the heuristic very simple to use and implement; on the other hand, it also means that there is little control of how difficult and rich the resulting neighborhood is. In practical implementations, an outer loop is added to the heuristic that adjusts the percentage of fixed variables in a dynamic fashion depending on the outcome of the sub-MIP solve. To date, RINS is probably the most effective of the LNS heuristics, and is implemented and routinely used by all MIP solvers, commercial and noncommercial.

Distance Induced Neighborhood Search A variation of RINS is the *distance induced neighborhood search* (DINS) heuristic [74] that, using tools from both local branching and RINS, can be considered as a sophisticated combination of soft- and hard-fixing. Essentially, DINS follows the principle that, ideally, it is worth looking for good solutions "close" to the solution of the LP relaxation of a node $\bar{x}$ because this is where solutions would have a better objective function value. This bias is taken into account by looking, for each $j \in \mathcal{I}$, to the distance $d_j = |\tilde{x}_j - \bar{x}_j|$. It follows that

---

[3] The approach used by local branching is often referred to as "soft-fixing."

- If $d_j \geqslant 0.5$, then changing $\tilde{x}_j$ might be a good idea in order to decrease the overall distance with respect to $\bar{x}$. Thus, DINS applies a so-called *rebounding* step by constraining the bounds of x_j in the sub-MIP so that the distance cannot increase.
- If $d_j < 0.5$, then keeping the variable at the value $\tilde{x}_j$ will not increase the distance to $\bar{x}_j$, so if $x_j \notin \mathcal{B}$, the variable is hard-fixed as well. For binary variables with small distance, both the previous integer feasible solutions and the solution of the LP relaxation at the root node are analyzed and if the values are in agreement the variable is hard-fixed. For all the binary variables that remain unfixed after this check, a local branching constraint is added to the sub-MIP.

The resulting sub-MIP is solved through a MIP solver call with work limits, and diversification can be applied naturally due to the changes of $\bar{x}$.

LB-RELAX Recently, another way of combining the RINS idea and the local branching paradigm was proposed [99]. LB-RELAX overcomes the issue that a local branching MIP tends to be the same size as the original MIP. It therefore only solves the initial LP of the local branching MIP and afterwards fixes all variables that coincide in the solution of this LP and the current incumbent solution. An alternative way of looking at this approach is to say that LB-RELAX creates a RINS sub-MIP, but instead of using an LP optimum of the original MIP, it uses the LP optimum of a local branching MIP as a reference point and applies a prescribed fixing rate.

Either way, in comparison to RINS, the approach sacrifices solution quality (from using the LP optimum) in favor of proximity to the incumbent (and therefore, hopefully, feasibility). In comparison to local branching, it significantly decreases the runtime (or increases the node throughput when assuming a given runtime) at the expense of pre-assigning some variables and thereby potentially missing solutions.

Relaxation Enforced Neighborhood Search The *relaxation enforced neighborhood search* (RENS) heuristic [19] is the first exception to the rule declared in the introduction of this chapter on using a feasible solution as a reference point. Indeed, RENS is an LNS primal heuristic, where a sub-MIP is defined by MIP operations and explored by calling a MIP solver in a black-box fashion, but it is used to find a feasible solution, not to improve it. More precisely, given a solution $\bar{x}$ of the LP relaxation of a MIP, RENS answers the question: what is the optimal rounding of $\bar{x}$? In other words, which is the optimal solution of the (binary) sub-MIP obtained by constraining $x_j \in \{\lfloor \bar{x}_j \rfloor, \lceil \bar{x}_j \rceil\}$, for each $j \in \mathcal{I}$?

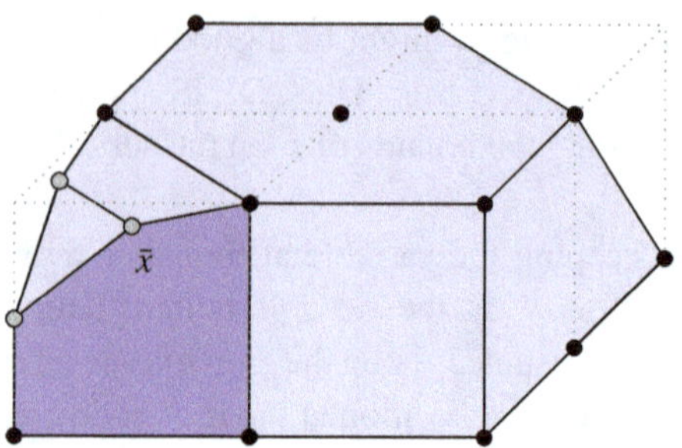

Figure 2.3 RENS sub-MIP on a small three-dimensional example. Fractional vertices are marked in gray, while integer vertices are marked in black. The $\bar{x}$ solution is the same as in Figure 2.2.

Note that such a sub-MIP might be infeasible, in which case we say that $\bar{x}$ is not "roundable." In the case of a binary MIP, that is, when $\mathcal{I} = \mathcal{B}$, the sub-MIP could even coincide with the original MIP if all binary variables were fractional in $\bar{x}$, so the RENS sub-MIP could be quite expensive in extreme cases. However, when a significant portion of general integer variables is present, or fewer fractional variables are present, the framework can be effective and, in principle, it can be applied to the solution of the LP relaxation at any node of the search tree.

A pictorial representation of the RENS branching sub-MIP on the same small three-dimensional example is given in Figure 2.3. Note that, by construction, given the same $\bar{x}$ solution and for any $\tilde{x}$ solution, the RENS sub-MIP is itself a sub-MIP of the RINS sub-MIP.

The computational experiments in the RENS paper [20] revealed that most MIPs feature LP optima that can be efficiently rounded. Surprisingly, the ability to round these solutions is not influenced by the fractionality of the LP solution. It also showed that rounding heuristics, as described in Chapter 3, often fail to find feasible solutions, even with roundable starting points, and rarely achieve optimal rounding.

2.3 Polishing Heuristic

The *polishing* heuristic [150] is a population-based LNS that maintains a fixed-size pool with the p best distinct solutions found during the branch-and-bound exploration, including the runs of MIP primal heuristics. The algorithm iterates over the following three steps:

(i) *selection*: it selects k solutions from the pool according to some criteria. Generally, $k = 2$ although sometimes in [150] all solutions are selected at the same time. The selected solutions are referred to – using an evolutionary algorithms terminology – as *parents*. The selection criterion when $k = 2$ is to pick a random solution from the entire pool and a second one, again at random but among those with a better objective function, so as to give a slight bias toward quality.

(ii) *combination*: the k solutions are combined by extending the RINS idea of comparing them and hard-fixing the variables whose values agree in all the k solutions. The resulting sub-MIP is then explored through a call to a MIP solver in a black-box fashion,[4] with some work limits. The obtained solution (if any) is called the *offspring* and is added to the pool if it is not a duplicate of another solution and, when the pool is full, if its value is smaller (i.e., better) than the worst solution in the pool. A pictorial representation of the combination sub-MIP on the usual three-dimensional example is given in Figure 2.4.

(iii) *mutation*: to avoid the solutions in the pool becoming too similar too quickly, d solutions are selected at random from the pool and a sub-MIP is built for each of them by hard-fixing a fraction f of randomly selected integer variables to their current value and leaving the remaining ones free. The resulting sub-MIPs are solved by calls of the MIP solver, and the obtained solutions are added following the previous scheme.

The above steps are repeated a predetermined number of times at each call of the polishing heuristic within the branch-and-bound algorithm. Indeed, the diversification is straightforward in the sense that the polishing algorithm through some randomness (of the mutation step) would change at any node of the MIP execution. It is easy to see that the scheme heavily depends on a number of parameters (e.g., k, d, f) and design choices (such as, for example, the k parents to be selected). Thus, the overall algorithm can be made more or less aggressive. Interestingly in that concern, CPLEX – the MIP solver the polishing algorithm was initially designed for – can be configured in such a way that, after a certain amount of time, it switches to a configuration where polishing runs continuously until the time limit is reached. This configuration is likely to be useful when the user's intuition is that solving the problem to optimality is out of reach, and concentrating on improving (or polishing, thus the name) the incumbent should be the focus.

---

[4] A similar combination scheme referred to as *crossover* has been independently proposed in [16].

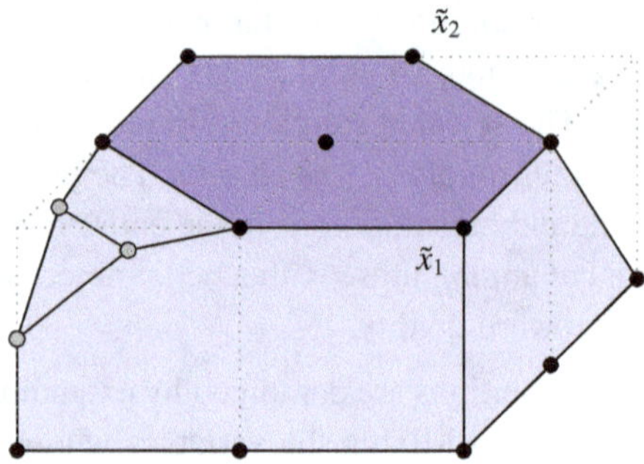

Figure 2.4 Combination sub-MIP on a small three-dimensional example. Fractional vertices are marked in gray, while integer vertices are marked in black.

Finally, it is worth mentioning that it is not necessary to wait until the pool has enough solutions to execute the algorithm. Indeed, the mutation step can be applied iteratively so as to produce any number of solutions starting from an initial one. Actually, the initial one can also be an infeasible or partial solution in the sense that it is enough to hard-fix a subset of variables and then rely on the sub-MIP to produce feasible solutions (mutation phase).

2.4 Exploiting Alternative Objective Functions

Following the discussion in [65], it is interesting to observe that the role of the objective function in MIP is two-fold. On the one hand, the objective function is needed/used to establish the measure of optimality for the feasible solutions. On the other hand, the objective function is responsible for guiding the search through the space of feasible solutions. Now, the latter role is clear when one needs to prove optimality (the "proof phase" according to [27]; see Section 1.3), while when the main goal is to improve the current incumbent ("improving phase") or find the first incumbent ("feasibility phase"), it is not totally obvious that the original objective function is the most suitable one. In fact, we will show in the following that it is not unusual to temporarily replace the original objective function with a more adequate one according to the scope of the search.

Zero-Objective Heuristic A common practice is for MIP solvers to temporarily remove the original objective function of a MIP P and solve the associated feasibility version of the problem, at least as a final attempt when other, cheaper heuristics have failed. Generally, this heuristic, referred to as the *zero-objective*, is run on a separate thread (when a MIP solver is executed in a multi-thread environment) until a first feasible solution has been found. Although the zero-objective is not properly based on the definition of a neighborhood, we present it here because a sub-MIP is still created and explored by the MIP solver used in a black-box fashion. The heuristic relies on the fact that, often, an empty objective function allows for more dual reductions during presolve; if this effect is not significant, a heuristic fixing can be applied in order to reduce the size of the sub-MIP further. Finally, as RENS presented in the previous section and some other heuristics presented later in the chapter, zero-objective is an LNS heuristic designed to find a feasible solution, and even more precisely in this case, zero-objective is used to find the *first* feasible solution. Indeed, differently from RENS and from other heuristics of this type, it only makes sense to call zero-objective once, that is, no diversification is applied.

Analytic Center Heuristic In the same spirit as the zero-objective heuristic, the *analytic center* (AC) heuristic [29] builds a sub-MIP and explores it through a call to a MIP solver in an attempt to find an initial incumbent. In particular, this primal heuristic is based on the interpretation of the analytic center of a bounded polyhedron (see Section 1.6.1, Eq. (1.8)) as a proxy for its *center of gravity*. Based on this interpretation, one can think of the value x_j^{ac} taken by a variable x_j at the analytic center as the expected value for such a variable. For a binary variable, this is equivalent to considering x_j^{ac} as the likelihood that x_j takes value 1 in a randomly sampled MIP solution. The AC heuristic follows this interpretation by replacing the original objective function of a MIP P by an auxiliary one whose coefficients are proportional to the analytic center solution values. In other words, the analytic center values are used as indicators of the direction variables are likely to move when going toward feasibility. The AC heuristic tends to be computationally expensive. First, the analytic center has to be computed. Second, the sub-MIP constructed by the heuristic has the same size as P. However, on the one hand, the analytic center computation is a byproduct of using an interior point method to solve the initial LP relaxation of P, which in the so-called concurrent mode[5] is currently the standard. On the other hand, the sub-MIP can be made smaller in a very natural way: by following the probability interpretation for a binary variable x_j, if $x_j^{ac} \geqslant \phi$ (resp., $x_j^{ac} \leqslant 1 - \phi$), then x_j can be fixed to 1 (resp., to 0), where ϕ is an appropriate threshold to guarantee the required reduction in size (and complexity) of the sub-MIP versus P.

---

[5] MIP solvers solve the initial LP relaxation of a MIP by running (in separate threads) primal simplex, dual simplex and the barrier interior point algorithm (plus crossover) until the first method proves optimality.

Proximity Search Different from the zero-objective and AC heuristics, the primal heuristic we consider here uses an initial feasible solution $\tilde{x}$ in the attempt to improve it. The *proximity search* heuristic [65] uses the Hamming distance $\Delta(x, \tilde{x})$ in the local branching constraint (2.1) to guide the search toward solutions that are close to $\tilde{x}$, but does it by replacing the original objective function of P by

$$\min \Delta(x, \tilde{x}). \tag{2.3}$$

Although the intent of using (2.3) is clear, the optimal solution of the sub-MIP obtained by swapping the objective functions in this way is trivially $\tilde{x}$ (whose distance is 0), so the constraints of P are amended by

$$c^T x \leqslant c^T \tilde{x} - \theta, \tag{2.4}$$

where $\theta > 0$ makes $\tilde{x}$ infeasible by requiring only solutions improving with respect to $\tilde{x}$ to be acceptable. Parameter θ somehow replaces parameter k in the local branching scheme: a larger θ value would allow faster convergence to quasi-optimal solutions at the potential risk of (i) complicating the search (i.e., going too far away from $\tilde{x}$), and (ii) failing, in case such an improvement is not possible. Although the proximity search heuristic can be clearly seen and has been presented in [65] as a standalone heuristic, its integration within a MIP solver (diversification) can be obtained similarly to that of local branching.

Local Branching Repair Local branching has been used in a different way with respect to that discussed in Section 2.1 in the attempt to build an initial feasible solution starting from a given infeasible one [63]. Given any mixed-integer *infeasible* solution $\hat{x}$, that is, a solution satisfying the integer requirements on the variables in $\mathcal{I}$, but violating some of the linear constraints $Ax \leqslant b$, a local branching scheme can be applied to repair $\hat{x}$ as follows:

(i) For each $i = 1, \ldots, m$ such that $\sum_{j \in \mathcal{N}} a_{ij} \hat{x}_j > b_i$,

    1. add an artificial variable $s_i \geqslant 0$;

    2. modify constraint i as $\sum_{j \in \mathcal{N}} a_{ij} x_j - s_i \leqslant b_i$;

    3. add a binary variable $y_i \in \{0, 1\}$;

    4. add the indicator constraint $s_i \leqslant M_i y_i$, where $M_i > 0$ is a big enough value, for example, $M_i = \sum_{j \in \mathcal{N}} a_{ij} \hat{x}_j - b_i$; and

    5. update the initially empty set of violated constraint V as $V = V \cup \{i\}$.

(ii) Solve the MIP updated with the variables and constraints in (i) by minimizing the artificial objective function $\phi = \sum_{i \in V} y_i$.

It is easy to see that solution $\hat{x}$ is feasible for the newly created MIP by setting all the y variables to 1. Then, solving the sub-MIP to optimality corresponds to finding a solution $\tilde{x}$ where $\phi = 0$, that is, no artificial variable s is needed. Thus, $\tilde{x}$ is feasible for the original MIP. Clearly, if $\phi > 0$ then the original MIP is infeasible. In [63], the initial solution $\hat{x}$ is obtained at the end of a phase in which the feasibility pump heuristic (see Chapter 4) is executed. However, any solution would do, ideally the closer to feasibility the better.

Alternating Criteria Search A significant extension of the local branching repair idea appeared in [134]. The method starts from an integer, but not necessarily linear feasible, initial reference point $\hat{x}_0$, and then alternates between two phases, each of which is based on solving a sub-MIP: one phase (called FMIP) aims at reducing the amount of infeasibility, and is based, in a way similar to local branching repair, on the idea of adding explicit slacks to the linear constraints and minimizing their sum in the objective function, like a MIP-equivalent of the phase-1 in the simplex method. The other phase (called OMIP) uses the original objective function but, notably, does not remove the artificial slacks: constraints can still be violated, but we have a *budget* constraint specifying that we cannot use more slack (in total) than we are using in the current reference solution $\hat{x}$. Loosely speaking, the method is allowed to redistribute constraint violation while trying to improve the objective function.

A key ingredient of the method is that, in both phases, a large subset of the variables is fixed to their value in $\hat{x}$; this hard fixing makes each sub-MIP much faster to solve than the local branching repair one. We refer to [134] for details on how to compute the initial solution and how to decide which variables to fix in each phase. We also note that the full heuristic in [134] makes extensive use of parallel computations: within each phase, several independent sub-MIPs are solved in parallel, and then their solutions are combined to generate the reference solution for the next phase.

2.5 Structure-Based LNS Heuristics

Formulating combinatorial optimization problems through MIP and solving them by MIP solvers has the clear advantage of leveraging decades of algorithmic and software development that led those solvers to become very effective and reliable. Sometimes, even more crucially, using MIP allows for fast deployment in industry by restricting the emphasis to modeling (with the MIP specs) instead of the time-consuming activity of developing special-purpose algorithms and codes. Because there is rarely a "free lunch," this comes with a cost, too. Namely, some of the potential knowledge of the structure of the problem at hand could get lost in the need to use the strict boundaries of the MIP language. In other words, such knowledge is hard (with very few exceptions) to communicate to a MIP solver, with the obvious consequence of not exploiting it for the solution process.

In this section, we discuss some structures that are either automatically recognized by MIP solvers or that the solvers can be instructed to recognize, and that can be effectively leveraged to design primal heuristics. More precisely:

- *Clique table.* Given the set $\mathcal{B} \subseteq \mathcal{I}$ of (indices of) binary variables, a clique $C \subseteq \mathcal{B}$ is a subset of variables that are pairwise incompatible, that is, they cannot take value 1 simultaneously, for all $i, j \in C, x_i + x_j \leqslant 1$. Then, the clique constraint

$$\sum_{j \in C} x_j \leqslant 1 \tag{2.5}$$

  is valid for P, and clearly provides rich pieces of information to the MIP process, for example, propagation and cutting planes.[6] MIP solvers inspect the MIP they receive on input and build a table of cliques by both recording those involved in constraints of type (2.5) and extending them through logical reasoning (see [1]).

- *Variable-bound graph.* Such a graph is constructed by the MIP solvers by inspecting constraints that involve precisely two variables, so as to record the relationships between their bounds. More precisely, a constraint $x + ay \leqslant b$, involving variables x and y with coefficients 1 and a, respectively, and right-hand side b, can be rewritten as $x \leqslant b - ay$, exposing that the upper bound of x is affected – depending on the sign of a – either by a variation of the *lower* bound of y, if $a > 0$, or by a change of the *upper* bound of y, if $a < 0$. This information is recorded through an arc in the variable-bound graph connecting the two nodes associated with the bounds of x and y. Of course, similar reasoning can be obtained by rewriting the constraint as $y \leqslant b - \frac{1}{a}x$, and the overall graph exposes the dependencies in a way that turns out to be suitable for propagation purposes.

- *Variable locks.* Down- and up-locks were formally introduced by Definition 1.5 and can be computed more straightforwardly with respect to the structures at the beginning of this section, as the MIP solver needs only to inspect the constraints once and count the number of negative and positive, respectively, coefficients of a variable in them. Variable locks provide a measure

---

[6] As soon as one variable belonging to a clique is fixed to 1, for example by branching, all the other variables in the clique can be fixed to 0.

of the effect of changing the bound of a variable in terms of the number of constraints that will be affected.

- *Permutation structure.* The feasibility space of several combinatorial optimization problems (the *quadratic assignment problem*, to cite only one) is defined by all possible permutations of a set of, say, n objects, the optimal solution being the permutation that minimizes a sometimes complex objective function. Permutations are easy to represent by a MIP through n^2 binary variables and $2n$ (linear) constraints [42]: given a square $n \times n$ matrix with the objects as rows and their positions as columns, it boils down to assigning each row to exactly one column and vice versa, which can be modeled by the constraints

$$\sum_{j=1}^{n} x_{ij} = 1, \quad i = 1, \ldots, n,$$

$$\sum_{i=1}^{n} x_{ij} = 1, \quad j = 1, \ldots, n,$$

over binary variables x_{ij}. Generally, however, MIP solvers are not instructed to recognize this structure,[7] thus clearly failing in using it even for producing an initial feasible solution that, in the case of a permutation, is trivial to construct.

These four structures have been used to devise effective primal heuristics. More precisely, the clique table, the variable-bound graph, and variable locks have been used in [69] as input to sophisticated variations of the simple fix-and-propagate LNS heuristic in Algorithm 3.

The basic scheme of the fix-and-propagate heuristic in Algorithm 3 receives on input the specific structure to be used, that is, the clique table, the variable-bound graph, or the variable locks, and iteratively tries to fix the variables by using the characteristic of the structure and exploiting its propagation logic. In case this (heuristic) fixing process leads to detect infeasibility, then backtracking is performed, until some condition is met. At the end of this phase, if the infeasibility has not been fixed (up to some work limits), then the algorithm exits without a solution. Otherwise, if at least a minimum number of variables (controlled by parameter α) have been fixed, an LP is solved to try to complete the solution, followed by a simple rounding step (see Section 3.1 for details).

---

[7] This is different from the previous three structures discussed that are automatically detected by MIP solvers. However, note that this automatic detection is clearly not granted. Simply, over the decades of MIP development, they were flagged as useful and general enough to drive the effort of writing special-purpose code to detect them. In other words, detecting new structures is a continuous work in progress that, as with many of the other ingredients of MIP solvers, is performed depending on the trade-off among factors such as generality and efficiency.

input : MIP $P(c, A, b, l, u, \mathcal{N}, \mathcal{I})$
input : global structure $\mathcal{S}$
input : size limit α
output : feasible solution or $NULL$, if no solution was found

1   $back = false,\ result = stop$
    /* try to fix all integer variables in structure $\mathcal{S}$ */
2   **while** $\{j \in \mathcal{I} \cap \mathcal{S} \mid l_j < u_j\} \neq \emptyset$ **do**
3     $(\widetilde{l}, \widetilde{u}) = (l, u)$
      /* get variable fixing based on global structure */
4     $(k, lower, result) = structureFixing(P(c, A, b, l, u, \mathcal{N}, \mathcal{I}), \mathcal{S})$
5   **if** $result \neq continue$ **then break**
      /* fix variable x_k */
6   **if** $lower$ **then** $\widetilde{u}_k = l_k$
7   **else** $\widetilde{l}_k = u_k$
      /* perform domain propagation */
8     $(P(c, A, b, \widetilde{l}, \widetilde{u}, \mathcal{N}, \mathcal{I}), inf) = domainPropagation(P(c, A, b, \widetilde{l}, \widetilde{u}, \mathcal{N}, \mathcal{I}))$
9   **if** inf **then**
        /* try to fix infeasibility by backtracking */
10       $(P(c, A, b, \widetilde{l}, \widetilde{u}, \mathcal{N}, \mathcal{I}), back, inf) = fixInf(P(c, A, b, \widetilde{l}, \widetilde{u}, \mathcal{N}, \mathcal{I}))$
11   **end**
12     $P(c, A, b, l, u, \mathcal{N}, \mathcal{I}) = P(c, A, b, \widetilde{l}, \widetilde{u}, \mathcal{N}, \mathcal{I})$
13   **if** inf **then return** $NULL$
14   **if** $back$ **then break**
15 **end**
    /* LP solving */
16 **if** $|\{j \in \mathcal{I} \mid l_j = u_j\}| \geqslant \alpha|\mathcal{I}|$ **then**
17     $\bar{x} = solve(P(c, A, b, l, u, \mathcal{N}, \emptyset))$
      /* round the LP solution */
18     $\hat{x} = simpleRounding(\bar{x})$
19   **if** $\hat{x}_j \in \mathbb{Z}\ \forall j \in \mathcal{I}$ **then return** $\hat{x}$
20   **else**
        /* LNS approach */
21       $\tilde{x} = solve(P(c, A, b, \widetilde{l}, \widetilde{u}, \mathcal{N}, \mathcal{I}))$ with working limits
22   **return** $\tilde{x}$
23   **end**
24 **else**
25   **return** $NULL$
26 **end**

Algorithm 3   Basic fix-and-propagate LNS.

If the rounding is not successful, a large neighborhood search is performed to try to produce a feasible solution.

The permutation structure has been used to devise an effective and fast local search heuristic in [152]. More precisely, in [152] an algorithm is implemented to detect at the end of the root node (and if some condition holds) if a permutation structure is present. In that case, the structure is used within an *iterated local search* [120] scheme that repeats the following steps:

 (i) perturb an initial solution;
 (ii) apply local search in the form of exploring a 2-opt neighborhood; and
(iii) determine whether or not to accept the new solution in accordance with some suitable scheme.

This scheme is implemented within the branch-and-bound tree through CPLEX callback functions (see [101] for details), and it is invoked at appropriate points of the search (see [152]). It is worth noting that this is the only heuristic presented in this chapter that does not explore the neighborhood by means of a call to a MIP solver. However, it has been presented in this chapter because it exploits some specific structure not directly captured by the MIP. In that sense, it provides an example of how users can exploit prior knowledge on their favorite combinatorial optimization problem to improve primal MIP capabilities.

2.6 Quiz

It's quiz time! This section is designed to test your understanding of the concepts covered in this chapter. Each question is followed by four possible answers, only one of which is correct. If you are unsure about some questions, you might find it helpful to work through the corresponding section again. Remember, sometimes the process of elimination can be as valuable a tool as direct problem-solving. By engaging with this quiz, you'll reinforce your knowledge and identify areas requiring further study. Take your time, think carefully and check the answers in the Appendix when you are done.

Question 1 What distinguishes MIP LNS heuristics from the more typical combinatorial local searches?

☐ the neighborhoods are guaranteed to be too expensive to be enumerated;
☐ the neighborhoods are defined in a problem-independent way, exploiting the MIP technology;
☐ they do not diversify;
☐ they do not randomize.

Question 2 The local branching constraint

- ☐ can only be stated for binary variables;
- ☐ reduces the complexity of the problem by fixing variables;
- ☐ makes the problem even harder to solve because of the additional constraint;
- ☐ restricts the search to solutions within distance k with respect to the ℓ_1-norm.

Question 3 The RINS sub-MIP

- ☐ always finds an improving solution if resource limits are set large enough;
- ☐ is always feasible;
- ☐ does not rely on branch and bound for diversification;
- ☐ can be run even before the root LP is solved.

Question 4 The RENS sub-MIP

- ☐ can be run even without an incumbent;
- ☐ is always feasible;
- ☐ relies on branch and bound for diversification;
- ☐ can be run even before the root LP is solved.

Question 5 The solution polishing heuristic

- ☐ needs at least two feasible solutions to be run;
- ☐ extends the RINS logic in its combination step;
- ☐ always picks the best two solutions in the pool;
- ☐ relies on branch and bound for diversification.

3

Rounding, Propagation and Diving

Rounding, propagation and diving heuristics are all relatively cheap heuristics whose purpose is to find a feasible solution $\tilde{x}$, often starting from a fractional solution $\bar{x}$. They share a common iterative nature, yet they differ in the amount of work and type of algorithms that are allowed at each step of the process. While these heuristics can make use of the knowledge of a feasible solution in their logic, they do not usually depend on it, so they can be run when no feasible solution is known. These heuristics are widely used in both commercial and open-source MIP solvers, although little literature is available on the implementation details for most of them (with some notable exceptions).

3.1 Rounding Heuristics

Rounding is a very natural approach to recover integrality from a fractional solution. However, in general no guarantees can be made on its effectiveness. Even in two dimensions, no feasible rounding of the optimal LP fractional solution $\bar{x}$ might exists; see, for example, Figure 3.1. In addition, on some models finding a feasible rounding is as complicated as solving the original problem. Take, for example, the Boolean satisfiability problem (SAT), where we have to look for an assignment to n Boolean variables $x_1, \ldots, x_n$ such that all the clauses in the problem evaluate to true. The problem can easily be formulated as an integer program: each clause gets translated into a $\geq$ linear inequality, and we have linear terms x_j (resp. $1 - x_j$) for each positive (resp. negative) literal in the clause, and the right-hand side is one. Under the reasonable assumption that each clause contains at least two literals (otherwise the problem can be trivially simplified), we have that setting all variables to $\frac{1}{2}$ is a feasible value for the LP relaxation; there are thus 2^n possible roundings, and deciding whether one of them is feasible is as hard as solving the original problem.

47

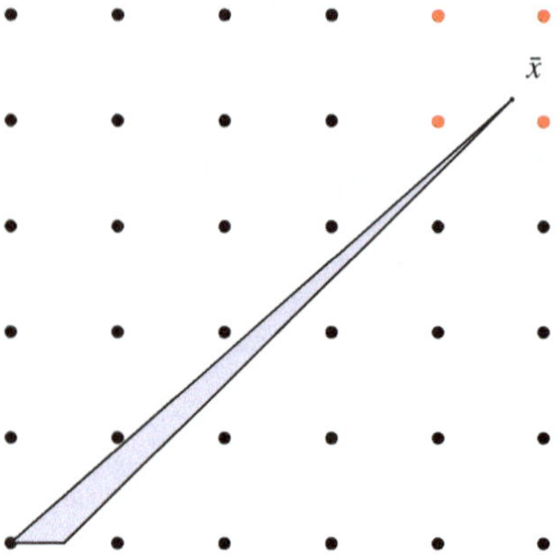

Figure 3.1 No feasible rounding.

The purpose of a rounding heuristic, given a fractional solution $\bar{x}$, is to round the fractional values $\bar{x}_j$, $j \in \mathcal{F}$, up or down in order to obtain a feasible integral solution $\tilde{x}$. Different rounding heuristics can be obtained depending on the order in which variables are considered for rounding, the logic to decide the rounding direction, and whether continuous and already integer variables are allowed to be modified in order to restore linear feasibility. As the number of possible combinations of these choices is quite large, we will just list the common options, plus some examples.

For choosing the order in which to round variables, the most common choices are:

- Natural order: use the variable index from left to right[1] in the problem formulation.

- Variable locks: round variables in order of increasing (or decreasing) variable locks.

- Objective/reduced costs/pseudocosts: use some objective-related quantity to sort variables.

- Other statistics based on the constraint matrix: the order might depend on how variables are linked in the constraints matrix. For example, a heuristic might want to round first variables that appear in many clique constraints.

- Random: this can be used as is, or as a tie breaker in conjunction with the previous strategies.

---

[1] We use "from left to right" to indicate a visualization of the model written as a MIP explicitly variable by variable.

As for choosing the rounding directions, popular options include:

- Rounding to the closest integer. While most natural, this rarely gives rise to feasible solutions [47].
- Rounding against the objective (i.e., round in the direction for which the objective gets worse). In [47], this is called *unfavorable rounding*.
- Rounding in the direction that preserves linear feasibility the most (e.g., where a variable lock is zero).

Of course, these strategies can be combined: for example, *gap rounding* [47] rounds near-integral values to the nearest integer and the remaining variables against the objective.

A basic rounding heuristic simply iterates over the set of fractional variables $\mathcal{F}$ from left to right, and rounds them into the direction that preserves linear feasibility the most; that is, it chooses the direction for which the corresponding lock is zero or for which the increase in cumulative constraint violation is smaller, or that even reduces violation. An even cheaper variant could also abort directly if a variable cannot be rounded in such a way as to preserve linear feasibility (this is called simple rounding in SCIP). A more elaborate, but still very cheap variant of simple rounding, is called ZI Round [171]. There, the fractionality of an LP-feasible solution is reduced by shifting fractional values toward integrality, but without necessarily rounding them, and always maintaining primal feasibility. For this reason, ZI Round might iterate over the set of fractional variables multiple times. Computational results showed that, on average, this is almost as fast as simple rounding, but is more successful.

A more elaborate version of the rounding heuristics just described, which is sometimes called *shifting*, allows for changing the values of continuous variables and/or integer variables with integer values in order to reduce constraint violation after a rounding step. Applied naïvely, this can easily lead to cycles, so care must be taken in the implementation in order to detect such cases and terminate. A variant of *shifting*, called integer shifting [1], is specifically tailored for models with continuous variables: it relaxes the continuous variables (by removing them), applies the basic shifting algorithm, reintroduces the continuous variables and solves a final LP to get the optimal values for them.

Finally, note that a rounding heuristic need not be applied only to an optimal LP relaxation. For example, simple rounding is cheap enough that it can be applied to intermediate LP solutions. In addition, a non-extremal solution, such as the *analytic center* [163], can also be used as input for rounding, as done, for example, in [138]. While more fractional, such a solution can sometimes be more easily rounded to a feasible integer solution because of its interior nature.

3.1.1 Inner Feasible Sets and Granularity

A completely different (and more principled) take on solution rounding is related to the concept of *inner feasible sets* [143]. The gist of the idea is that while rounding a solution coming from the standard LP relaxation might be nontrivial, we can define a different set from which any rounding is actually feasible. For simplicity, let us consider a pure integer linear program (thus, with no continuous variables),[2]

$$
\begin{aligned}
\min \quad & c^{\mathsf{T}} x \\
\text{s.t.} \quad & a_i^{\mathsf{T}} x \leqslant b_i, \quad \text{for all } i \in \{1, \ldots, m\}, \\
& l_j \leqslant x_j \leqslant u_j, \quad \text{for all } j \in \{1, \ldots, n\}, \\
& x \in \mathbb{Z}^n,
\end{aligned}
$$

and denote by F the feasible region of its LP relaxation. The *inner feasible set* F^- associated with F is formally defined as

$$
F^- = \{x \in \mathbb{R}^n : x + K \in F\},
$$

with $K = B_\infty(0, \frac{1}{2}) = \{x \in \mathbb{R}^n : \|x\|_\infty \leqslant \frac{1}{2}\}$. The intuition is that, given an arbitrary point in F^-, any of its roundings is guaranteed to be in F, and thus be a feasible solution. In order to make this definition usable in an algorithm, we need to define F^- in closed form. It was shown in [143] that this is easily obtained as

$$
F^- = \left\{ x \in \mathbb{R}^n : \quad
\begin{aligned}
& a_i^{\mathsf{T}} x \leqslant b_i - \frac{1}{2}\|a_i\|_1 \quad \text{for all } i \in \{1, \ldots, m\}, \\
& l_j + \frac{1}{2} \leqslant x_j \leqslant u_j - \frac{1}{2} \qquad \text{for all } j \in \{1, \ldots, n\}
\end{aligned}
\right\}.
$$

So a possible heuristic strategy is to find a point in F^-, for example, by solving the corresponding LP, and then, if this LP is not infeasible, rounding its optimal solution. Unfortunately, as defined above, the inner feasible is very likely to be empty for many practical MIPs: note, for example, that any binary variable would be fixed to $\frac{1}{2}$ by the construction above, which is unlikely to be feasible for all constraints. Also, assuming equality constraints are present and modeled as pairs of inequalities, the inner feasible set is guaranteed to be empty in the form presented above.

To circumvent those obvious drawbacks, in [143] it was proposed to consider an enlarged set $F_\delta \supseteq F$, while still maintaining the property that $F_\delta \cap \mathbb{Z}^n$

---

[2] The concepts described in this section are not restricted to the pure integer case; we are making this assumption just to simplify notation. The reader is referred to [143] for the general case.

defines the feasible set of the integer program. Assuming, without loss of generality, that all the problem data is integer, and that all coefficients in a row are relatively prime, F_δ can be obtained in closed form as:

$$F_\delta = \left\{ x \in \mathbb{R}^n : \quad a_i^\mathsf{T} x \leqslant b_i + \delta \quad \text{for all } i \in \{1, \dots, m\}, \right.$$
$$\left. l_j - \delta \leqslant x_j \leqslant u_j + \delta \quad \text{for all } j \in \{1, \dots, n\} \right\}$$

for any $\delta \in (0, 1)$. We can apply the inner parallel set construction to the enlarged set F_δ, and finally obtain:

$$F_\delta^- = \left\{ x \in \mathbb{R}^n : \quad a_i^\mathsf{T} x \leqslant b_i + \delta - \frac{1}{2}\|a_i\|_1 \quad \text{for all } i \in \{1, \dots, m\}, \right.$$
$$\left. l_j + \left(\frac{1}{2} - \delta\right) \leqslant x_j \leqslant u_j + \left(\delta - \frac{1}{2}\right) \quad \text{for all } j \in \{1, \dots, n\} \right\}.$$

An integer program is called *granular* if its set F_δ^- is non-empty. Thus, for granular integer programs, we can always find a feasible solution in polynomial time by solving an LP over F_δ^- and then rounding to the nearest integer. In a follow-up publication, a diving framework (see Section 3.3) based on the concepts above was presented [142].

3.1.2 Solution Repair

The outcome of a rounding heuristic is often a complete yet infeasible solution. Instead of just throwing it away, an alternative option is to enter a so-called *solution repair* step, whose purpose is to modify the complete solution iteratively, trying to reduce the number of constraint violations until a feasible solution is found. The most prominent example of this approach is the WalkSAT [156] heuristic for Boolean satisfiability problems: a complete assignment is generated at random and then, at each iteration, a violated clause is selected at random and a literal in it is flipped, using a simple logic that is a careful mix of greedy and random walk behavior (hence the name) until a feasible solution is found or some resource limits are hit.

Similar approaches have also been proposed in the MIP setting: the general idea is to start from an infeasible solution (either at random, or obtained by rounding), replace the original objective (if any) with a new function evaluating the cumulative constraints' violation, and perform some form of local search, trying to minimize the total violation. The type of local search depends on the method and can either be combinatorial, that is, based on moves like flips and swaps, or based on the sub-MIP paradigm. Examples of the first type (combinatorial) are the *LocalSolver* approach [14, 115] and the recently proposed *feasibility jump* [121] and *Local-ILP* [112]. The sub-MIP paradigm is used in the local branching repair approach and in the *alternating criteria search* method from Chapter 2.

3.2 Propagation Heuristics

In rounding heuristics, the constraint matrix plays only a very limited role: at most, it is used to compute a cumulative constraint violation score that is used to evaluate in which direction to perform the next rounding operation, or to stop.

A natural improvement, which leads to propagation heuristics, consists of performing constraint propagation, also called bound strengthening, after each rounding, in a scheme often referred to as *fix-and-propagate* [6, 25, 69, 128]. Constraint propagation might directly fix some additional variables, so that we do not have to heuristically guess a rounding direction for them, or even detect infeasibility earlier, so that we do not have to waste time completing a rounding that cannot lead to a feasible solution. The price to pay for this is an increased computational effort, as constraint propagation can be expensive, yet it is usually cheap compared to other techniques like solving LPs. Notice also that constraint propagation makes the order in which we decide to round variables more important.

Propagation heuristics can be seen as a bridge between rounding, described in Section 3.1, and diving, which will be described in Section 3.3. Indeed, we could describe a propagation heuristic as a diving heuristic in which we skip solving the LPs. Given that constraint propagation is expensive, and rounding directions are only heuristically chosen, some limited amount of backtracking is usually allowed in a propagation heuristic. A very common case is to allow a one-level backtrack: if constraint propagation detects infeasibility, the latest rounding is undone and the opposite direction is taken, from which the heuristic continues as usual. Note that one-level backtrack can be implemented quite efficiently and it is already available in MIP solver to implement *probing* [154], a form of preprocessing where a set of binary variables is temporarily fixed (i.e., *probed*) and constraint propagation is performed to derive the implications of those fixings. The implications are then analyzed to derive globally valid constraints for the problem, such as fixings, tighter bounds and aggregations.

A general scheme for a propagation heuristic is outlined in Algorithm 4.[3]

---

[3] In Algorithm 4, with a slight abuse of notation, we are using $\tilde{x}$ to denote a vector that we are trying to transform from LP feasible to integer feasible. If the process is unsuccessful, no solution is returned.

```
1  x̃ = x̄
2  Q = I
3  while Q ≠ ∅ do
4  │   select j ∈ Q
5  │   choose v_j ∈ D_j
6  │   propagate D_j = {v_j}
7  │   if propagation detects infeasibility then
8  │   │   apply one-level backtrack
9  │   else
10 │   │   x̃_j = v_j
11 │   │   Q = Q \ {j}
12 │   end
13 end
14 if x̃ is feasible for P then return x̃
15 return NULL
```

Algorithm 4   Basic fix-and-propagate scheme.

As in the case of rounding heuristics, there is a very large range of possibilities as far as variable/value strategies are concerned. Those listed in Section 3.1 are as relevant for fix-and-propagate as they were for rounding, but there are more elaborate strategies in the literature. The best example is probably *shift-and-propagate* [25]. There, a complete yet infeasible integer reference solution is used to decide the variable ordering, with variables sorted by the nonincreasing number of violated constraints they appear in. Then, within the fix-and-propagate scheme, the same reference solution, which is kept up to date with respect to the current partial assignment, is used to compute the tentative value to fix the next variable too. In detail, shift-and-propagate starts with a reference solution in which all variables are set to their lower bound (which can be assumed to be zero, after an appropriate normalization), and variables are ranked according to the number of violated constraints with respect to this reference solution. At each iteration, an unfixed variable x_j is chosen and a fixing value t_j inside its domain is calculated as the value giving the best reduction in violation count with respect to the current reference solution. Constraint propagation is then performed; if infeasibility is detected, one-level backtracking is performed, otherwise the fixing is kept and the reference assignment is updated with the new value t_j for x_j and possibly updated values for the variables affected by the latest fixing (because of constraint propagation). After the reference solution is updated, constraint violations are reevaluated as well to get an updated ranking of unfixed variables in order to be ready for the next iteration. Additional variable/value strategies are computationally evaluated in [153]. There, the basic *fix-and-propagate* is also amended with a *repair* step, loosely inspired by WalkSAT [156], whose purpose is to try to fix the main weaknesses of the basic approach, namely the inability to recover quickly from past mistakes and the strong dependence on the variable and value selection strategies. The computational results in [153] showed that the resulting methods achieve a higher success rate in finding an initial feasible solution on a heterogeneous test set of models.

It is worth noting that it is also possible to relax the one-level backtrack rule and allow for unlimited backtracking, effectively turning the fix-and-propagate heuristic into a depth-first exhaustive search. While this is clearly not an effective option as far as solving the original problem is concerned, as it would be equivalent to perform branch and bound without using LP relaxations, with proper limits it might still be a useful heuristic. The intuition here is that a propagation-based enumeration can process nodes much faster than regular branch and bound, is usually able to explore thousands of nodes in a fraction of the time needed to solve the root LP relaxation, and could thus find primal solutions that are deep in the tree. Exploiting such a propagation-based search to collect primal solutions (and potentially other useful pieces of information, e.g., conflicts) was proposed and computationally evaluated in [21], and subsequently extended in [31]. Finally, we note that fix-and-propagate heuristics can also play the "preprocessing role" in other heuristics. We have already seen an example of this in the structure-based LNS heuristics in Chapter 2, where the clique and implication tables were used in a fix-and-propagate fashion, and a sub-MIP was solved only on the remaining variables.

3.3 Diving Heuristics

Diving heuristics are the logical next step to propagation heuristics, in which after fixing a variable by rounding we not only apply constraint propagation, but also resolve the LP relaxation. Again, the procedure is iterated until either infeasibility or cutoff is detected (by constraint propagation or the LP) or a integral solution is found. Note that in diving the only change between one LP and the next to be solved is some bound tightening; hence diving exploits the warm-starting capabilities of the dual simplex method quite effectively. A general scheme for a diving heuristic is outlined in Algorithm 5.

Diving heuristics simulate an incomplete branch-and-bound search, in which only one child node is created at each branching decision, and that *quickly goes*

```
 1  F = {j ∈ I | x̄_j ∉ ℤ}
 2  while F ≠ ∅ do
 3  │   select j ∈ F
 4  │   tighten x_j ≤ ⌊x̄_j⌋ or x_j ≥ ⌈x̄_j⌉
 5  │   propagate and resolve LP
 6  │   if infeasibility detected then
 7  │   │   apply one-level backtrack (and resolve LP)
 8  │   end
 9  │   if still infeasible then
10  │   │   return NULL
11  │   end
12  │   F = {j ∈ I | x̄_j ∉ ℤ}
13  end
14  return x̄
```

Algorithm 5   Basic diving scheme.

down in the tree, hence the name. While branch and bound would eventually, if needed, find those solutions anyway, factoring out these dives into dedicated heuristics has at least the following double benefit:

- It allows us to find the corresponding integer solutions much earlier than regular branch and bound would.

- It allows for using different variable/value ordering strategies that are more suitable for finding primal solutions quickly. Intuitively, we do not want to *spoil* the branch-and-bound tree with decisions that are good for finding solutions but that are potentially disastrous for proving optimality, the main task of a complete search.

A few diving strategies are described in [1, 16]. They differ on how the next variable to fix is picked, and to which value. Namely:

- Fractional diving: a variable with minimal fractionality is chosen and fixed to the nearest integer.

- Coefficient diving: a variable x_j minimizing the expression based on variable locks $\min\{\overline{\kappa}_j, \underline{\kappa}_j\}$ is chosen and rounded in the direction of the smallest lock number. The distance between the fractional and rounded value is used as a tie breaker.

- Line search diving: this is based on a node selection idea in [126]. Given a variable x_j, line search diving compares its current fractional value $\bar{x}_j$ with the value $\bar{x}_j^R$ it had at the root node, and rounds it up (resp. down) if it is higher (resp. lower). The geometric intuition is to connect the root node LP solution $\bar{x}^R$ to $\bar{x}$ by a line, and extend this line past $\bar{x}$ until it hits a integer value for one of the variables in $\mathcal{F}$. This also gives a strategy for deciding which variable to pick, namely the first one that becomes integral.

- Pseudocost diving: this strategy uses a combination of line search diving, fractionality and pseudocosts in order to decide in which direction to round a fractional variable x_j, $j \in \mathcal{F}$. A score is calculated for each variable: the fractionality is multiplied by the quotient between the pseudocosts of the variable to the direction where it should not be bounded and the pseudocosts of the variable to the direction where it should be bounded. A low score indicates a direction with high decrease in the objective function, a high score a direction with small decrease. A variable with highest score will be chosen for bounding.

- Vectorlength diving: this strategy is tailored toward set covering and set partitioning models, a common (sub)structure in applications. The idea is to choose a variable that would cover the largest number of constraints with the smallest objective cost incurred by rounding. The rounding is always against the objective.

An additional strategy is the so-called guided diving. This strategy, introduced in [51], requires the knowledge of an integer feasible solution, and thus can be considered an improving heuristic. In the original paper, this strategy was applied within the main search tree, and it consists of picking a variable with the current branching rule and then choosing first the child node in which the variable is moved into the direction of the current incumbent $\tilde{x}$. However, it can easily be implemented outside the tree as with any other diving heuristic, with a suitable variable selection strategy; SCIP for example implements it this way, always choosing the variable whose fractional value is closest to its value in the incumbent.

Another list of diving strategies is described in [146], all based on the set of currently *active* constraints, that is, those that are tight at the current fractional solution. The idea is to assign a score w_{ij} to each pair (i, j), where i is the index of an active constraint and j the index of a fractional variable, according to some strategy (the paper lists seven). The strategies are all based on the intuition that one would prefer a variable that influences active constraints the most, and each strategy follows this intuition by using a different measure. The variable with the highest total weight is eventually chosen as the next variable to dive on. The approach was later extended in [147] to take (approximate) probabilities for satisfaction of constraints.

Diving strategies have also been extended to constructs that are commonly supported as an extension to the basic MIP framework, such as semicontinuous variables or special ordered sets. A recently proposed strategy that deals with indicator constraints encoding semicontinuous variables is *indicator diving* [86].

Another interesting approach is the one proposed in [173]; there, diving strategies were proposed that target the generation of valid conflicts. The intuition is that if those dives succeed, then the feasible solution is expected to be of very good quality, otherwise we can still derive a valid conflict constraint to be used later in the search.

Finally, a first attempt at diving by imposing general constraints (as opposed to simple bound tightenings) was proposed [122]. While more expensive in terms of LP solving and in terms of choosing which general constraint to add next, if used with care the approach can lead to significantly shorter dives, and thus more effective heuristics overall.

3.4 Quiz

It's quiz time! This section is designed to test your understanding of the concepts covered in this chapter. Each question is followed by four possible answers, only one of which is correct. If you are unsure about some questions, you might find it helpful to work through the corresponding section again. Remember, sometimes the process of elimination can be as valuable a tool as direct problem-solving. By engaging with this quiz, you'll reinforce your knowledge and identify areas requiring further study. Take your time, think carefully and check the answers in the Appendix when you are done.

Question 1 For a general MIP, rounding the fractional solution associated with the root LP relaxation

- ☐ is guaranteed to yield the optimal integer solution;
- ☐ is guaranteed to yield a near-optimal integer solution;
- ☐ is guaranteed to yield a feasible solution;
- ☐ might not give a feasible solution.

Question 2 Let us consider the fractional solution of a binary knapsack problem in normalized form (all positive data and no single item weight exceeding the knapsack capacity). Which of the following variable/value strategies is guaranteed to yield a feasible solution?

- ☐ natural order/toward the objective;

- ☐ increasing variable locks/nearest integer;
- ☐ random/against the objective;
- ☐ natural order/nearest integer.

Question 3 Let us consider the binary knapsack problem of the previous question, but in the context of fix-and-propagate. Which variable/value strategies are now guaranteed to yield a feasible solution?

- ☐ (only) the same as in Question 2;
- ☐ none;
- ☐ all of them, assuming one-level backtracking;
- ☐ all of them, even without backtracking.

Question 4 Diving heuristics are used inside a branch-and-bound algorithm

- ☐ because branch and bound will not find the optimal solution otherwise;
- ☐ so as to not spoil the search with branching decisions unsuited for the optimality proof;
- ☐ because we are solving LPs anyway, so they have no overhead;
- ☐ because they are guaranteed to improve performance.

4

The Feasibility Pump Family

The fundamental idea of all feasibility pump algorithms is to construct two sequences of points that hopefully converge to a feasible solution of a given optimization problem. One sequence consists of points that are feasible for a continuous relaxation (e.g., the linear programming relaxation of a MIP), but possibly integer infeasible. The other sequence consists of points that are integral, but might violate some of the constraints. The next point of one sequence is always generated by minimizing the distance to the last point of the other sequence, possibly by using different distance measures in either case (e.g., the ℓ_1- or ℓ_2-norm).

4.1 Basic Algorithm

The feasibility pump[1] (FP) was originally introduced in 2005 [61] for mixed-integer 0-1 programs, that is, for the special case of MIPs in which $l_j = 0$ and $u_j = 1$ for all $j \in I$. The main idea is as follows. The LP relaxation of a MIP is solved. The LP optimum $\bar{x}$ is then rounded to the closest integral point,

$$\tilde{x} = \left\{ \begin{array}{ll} [\bar{x}_j] & \text{if } j \in I, \\ \bar{x}_j & \text{if } j \notin I, \end{array} \right. \tag{4.1}$$

where $[\cdot]$ represents scalar rounding to the nearest integer and, with the slight abuse of notation already used in Chapter 3, $\tilde{x}$ represents a mixed-integer vector not necessarily LP feasible. This part of the FP algorithm is called the *rounding step*. If $\tilde{x}$ is not feasible for the linear constraints, the objective function of the LP is changed to the ℓ_1-norm distance function

---

[†] This chapter is based on a previously published article. Used with permission of Elsevier. [28]
[1] The name of the heuristic was coined as a tribute to the "electron pump" in Isaac Asimov's novel *The Gods Themselves*.

59

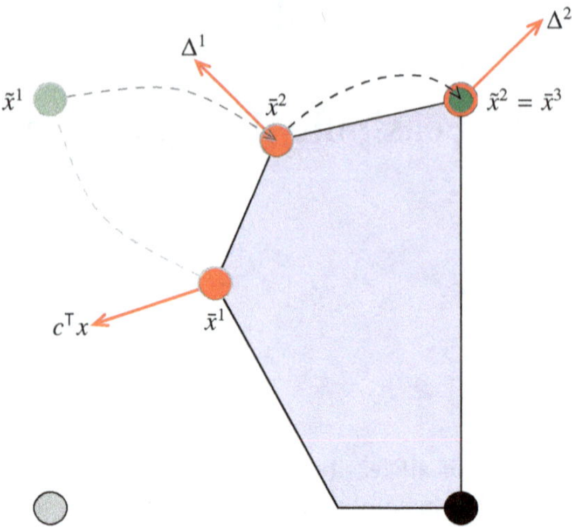

Figure 4.1 Feasibility pump for MIP, sequences of constraint-feasible points (red), integer feasible points (green), original objective (dotted red) and distance functions (solid red). Used with permission of Elsevier Science & Technology Journals, from [28]; permission conveyed through Copyright Clearance Center, Inc.

$$\Delta(x, \tilde{x}) := \sum_{j \in I} |x_j - \tilde{x}_j| = \sum_{j \in I \,:\, \tilde{x}_j = 0} x_j + \sum_{j \in I \,:\, \tilde{x}_j = 1} (1 - x_j), \qquad (4.2)$$

and a new LP point $\bar{x}$ is obtained by minimizing $\Delta(x, \tilde{x})$ over the linear constraints of the MIP. The process is iterated until $\tilde{x} = \bar{x}$, which implies feasibility (with respect to the MIP). The operation of obtaining a new $\bar{x}$ from $\tilde{x}$ is known as the *projection step*, as it consists of projecting $\tilde{x}$ to the feasible set of a continuous relaxation of the MIP along the direction $\Delta(x, \tilde{x})$. Two iterations of the algorithm are illustrated for a simple example in Figure 4.1, while a pseudocode description of the method is given in Algorithm 6.

The algorithm thus produces two sequences $\{\bar{x}^k\}_{k=1}^K$ and $\{\tilde{x}^k\}_{k=1}^K$ for a finite K, which is either the iteration at which a feasible solution for (1.1) is found or some limit set to guarantee termination. All points of the sequence $\bar{x}^k$, with k denoting the iteration count of the FP, are feasible for the LP relaxation; all points $\tilde{x}^k$ are integral, that is, $\tilde{x}_j^k \in \mathbb{Z}$ for all $j \in I$. Thus, $\tilde{x}^k = \bar{x}^k$ implies integrality and constraint-feasibility, which means that the corresponding point is feasible for the MIP.

The main obstacle for the original FP algorithm (and many of its successors) is *cycling*: after some iterations, it may happen that $\tilde{x}^k = \tilde{x}^{k'}$ with $1 \leqslant k' < k$. In this case, the procedure would enter a loop, revisiting the sequence $\tilde{x}^{k'} \cdots \tilde{x}^{k-1}$

---

input : MIP $\equiv \min\{c^T x : x \in P,\ x_j \text{ integer } \forall j \in I\}$
output: a feasible MIP solution $\bar{x}$ (if found)

1    $\bar{x} = \arg\min\{c^T x : x \in P\}$
2   **while** *not termination condition* **do**
3   **if** $\bar{x}$ *is integer* **then return** $\bar{x}$
4       $\tilde{x} = \mathtt{Round}\,(\bar{x})$
5   **if** *cycle detected* **then** $\mathtt{Perturb}\,(\tilde{x})$
6       $\bar{x} = \mathtt{LinearProj}\,(\tilde{x})$
7 **end**

Algorithm 6   Feasibility pump – the basic scheme.

(and $\bar{x}^{k'} \cdots \bar{x}^{k-1}$) over and over again. Since the central idea of FP is to bring the sequences close together, the risk of cycling is "naturally encoded" in the procedure and occurs very frequently in computational experiments. In the original FP, this issue is handled via a simple random perturbation: some of the variables in $\tilde{x}^k$ are flipped to the other bound before continuing the procedure. If the issue persists, a more aggressive perturbation akin to a *restart* is performed.

Figure 4.2 shows the different behavior of FP on two sample instances: while in Figure 4.2(a) the distance between $\bar{x}$ and $\tilde{x}$ is rapidly brought to zero without the need of any perturbation/restart, in Figure 4.2(b) FP exhibits a much less satisfactory behavior, with frequent restarts and perturbations that yield large oscillations of the distance function and hence reduces the probability of success of the method. It is a crucial component of many FP extensions that cycling is addressed directly, made more unlikely, or avoided completely; see more details in Section 4.4.

It was demonstrated in [61] that the FP is very effective in finding feasible solutions, but these are often of low quality – not surprising considering the fact that the original objective is only considered in the very first iteration. The authors suggest using subsequent runs of the FP to get better solutions. After each successful run, a primal bound constraint $c^\mathsf{T} x \leqslant \beta c^\mathsf{T} \bar{x} + (1 - \beta)c^\mathsf{T} \tilde{x}$ is added to the MIP, with $\beta \in (0, 1)$, $\bar{x}$ being an optimal solution of the original LP relaxation, and $\tilde{x}$ being the solution from the previous FP run.

In [15], an FP variant for mixed-integer programs with general integer variables was introduced. The authors used an auxiliary variable d_j and two auxiliary constraints for each general integer variable x_j to represent the two linear

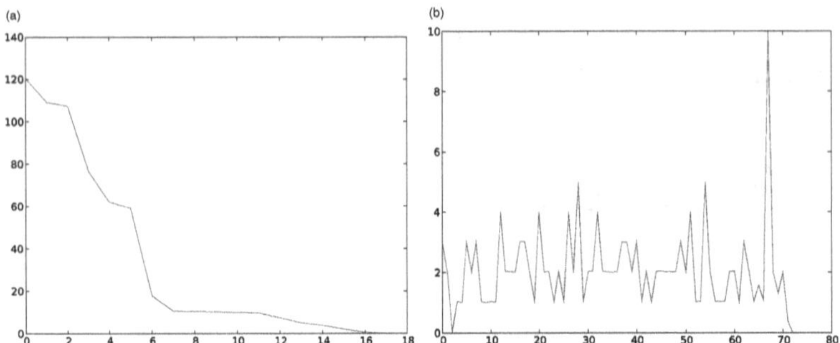

Figure 4.2 Two very different FP behaviors (integrality distance vs. iterations): (a) good, (b) bad. Used with permission of Elsevier Science & Technology Journals, from [28]; permission conveyed through Copyright Clearance Center, Inc.

pieces of the absolute values $|x_j - \tilde{x}_j|$ whenever a variable x_j is not sitting at one of its bounds in $\tilde{x}_j$. Then the objective function for the projection step is a modified version of function (4.2), namely

$$\Delta(x, \tilde{x}) := \sum_{j \in I \,:\, \tilde{x}_j = l_j} (x_j - l_j) + \sum_{j \,:\, \tilde{x}_j = u_j} (u_j - x_j) + \sum_{j \in I \,:\, l_j < \tilde{x}_j < u_j} d_j,$$

with $d_j \geqslant x_j - \tilde{x}_j$ and $d_j \geqslant \tilde{x}_j - x_j$ for all $j \in I$ with $l_j < \tilde{x}_j < u_j$.

Further, the authors suggested splitting the FP procedure into different stages. In stage I, the ℓ_1-norm objective function is defined only on the binary variables, if any. The auxiliary variables and constraints are only used in stage II, which kicks in if the solution is mixed 0-1 feasible (or work limit conditions have been reached), and also considers distances on general integers. Finally, stage III is an enumeration phase consisting of a truncated MIP search. For this, the objective function of the original MIP is replaced by the distance with respect to the point $\tilde{x}^k$ from the previous stages that was closest to the LP relaxation. A similar definition of the objective function was later used in [65] in the proximity search algorithm. A different formulation for stage III, recently proposed in [127], is based on the LP feasible points generated in the previous stages: those points are collected in a set $\bar{X}$ and the bounds of the integer variables x_j for all $j \in I$ for the stage III MIP are computed as $\lfloor \min_{\bar{x} \in \bar{X}} \bar{x}_j \rfloor$ and $\lceil \max_{\bar{x} \in \bar{X}} \bar{x}_j \rceil$, basically extending the RENS [19] logic to multiple reference vectors.

The main idea of the FP is to decompose the original problem into two problems, retaining linear and integrality constraints. Many of the extensions described in this chapter aim at solving each of the decomposed problems with an eye on the constraints of the other problem, in an attempt to accelerate convergence.

Besides enhancements of the FP for MIP itself, a natural direction of investigation is the extension of the FP idea to other contexts and/or to broader problem classes. In [3, 4], a FP-like algorithm, called *PumpReduce*, was used to generate alternative LP optima that can be used for improved cut generation and filtering. According to [29], XPRESS has a similar algorithm. In [30], a similar algorithm was used as a basis for a branching scheme, called cloud branching, which is based on a set of relaxation optima. Cloud branching applies a FP-like objective to the optimal face of a MIP's LP relaxation to compute multiple alternative optima, typically with reduced fractionality. These alternative optima are then used to filter potential candidate variables for which traditional branching rules such as strong branching are performed. An extension in [145] dealt with bi-objective pure integer linear programming, where FP and local search heuristics are designed to approximately generate the so-called non-dominated frontier.

4.2 Improving the Rounding Step

It was observed in [66] that rounding a variable can be interpreted as a temporary fixing. The authors suggested propagating the minimum and maximum activities of linear constraints using these fixings as local bounds. The propagation is done via the well-known bound strengthening techniques described in Chapter 1. By this, variables might be fixed to a value that is not the closest integer (compared to $\bar{x}^k$), in order to avoid a linear constraint becoming infeasible with respect to the current temporary fixings. Thus, the rounded solution might be further away from the last LP optimum than for the original FP, but it will be closer to the feasible region. In their computational experiments, the authors demonstrated that the so-called feasibility pump 2.0 needs fewer iterations and produces slightly better solutions.

Several authors [8, 34] introduced variants of the FP that use integral reference points $\tilde{x}$ that are closer to the interior of the LP polyhedron. Therefore, both publications suggest connecting the LP optimum $\bar{x}$ with the analytic center (1.8) of the LP and then performing a search for integer points that are roundings of points on that line segment.

In the algorithm presented in [8], points are sampled on the line segment $\bar{x} - x^{\mathrm{ac}}$. Each of those sampled points is rounded and tested for feasibility. If none of the rounded points are feasible, a new integral reference point $\tilde{x}$ is chosen as the rounded point that minimizes the ℓ_∞-distance to the line segment point it has been rounded from. In [34], the procedure was extended by several innovative ideas. First, it is proven that the set of all integral points that are roundings of some point of the line segment can be computed very efficiently. This improves the sampling step. The main overhead of the procedure in [8] lies in the computation of the analytic center in order to get a direction pointing from the LP optimum toward the interior of the polyhedron. The algorithm of [34] uses a conic combination of the normal vectors of all constraints violated by $\tilde{x}$ as a cheap heuristic approximation for a ray pointing toward the center. At the other extreme, [138] proposed a version of the FP with analytic centers that additionally applies a *recursive central rounding* procedure, which iteratively fixes some of the integer variables and recomputes the analytic center.

4.3 Improving the Projection Step

The main direction of modification for the projection part of the FP was the use of different objective functions for the LP. In [5], a simple trick was shown to overcome a main weakness of the FP: despite success in many instances, the produced solutions are often of poor quality. This seems natural, since the original objective of the MIP is only considered when computing the first LP solution $\bar{x}^1$. Other from that, solution quality does not play a role, neither for the rounding step nor for the projection step. The authors of [5] therefore suggested replacing function (4.2) by a convex combination of (4.2) and the original objective $c^\mathsf{T}x$, namely

$$\Delta_\alpha(x, \tilde{x}) := (1 - \alpha)\Delta(x, \tilde{x}) + \alpha\frac{\sqrt{|I|}}{\|c\|}c^\mathsf{T}x,$$

with $\alpha \in [0, 1]$. Here, $\| \cdot \|$ is the Euclidean norm of a vector. The convex combination factor α, and hence the influence of $c^\mathsf{T}x$, is reduced in every iteration. As a nice side effect, this often enables the algorithm, called the objective feasibility pump, to avoid cycling since the objective function $\Delta_\alpha(x, \tilde{x})$ depends on the iteration count and will be different even when the same point $\tilde{x}$ is visited more than once. As a result, the objective FP finds solutions of better quality on average. However, this comes at the price of a slightly reduced success rate and increased number of iterations/runtime. The norm of the objective vector is used to try to bring the two terms of the convex combination to a similar scale. Recently, a more effective way to obtain the same effect was given in [127]: the idea is to scale both terms by a term that depends on the previous LP solution in the sequence. In detail, the new objective function for the projection step at iteration k reads:

$$\Delta_\alpha^k(x, \tilde{x}^k) := (1 - \alpha)\Delta(x, \tilde{x}^k) + \alpha\frac{|\Delta(\bar{x}^{k-1}, \tilde{x}^k)|}{|c^\mathsf{T}\bar{x}^{k-1}|}c^\mathsf{T}x.$$

In [57], the FP was interpreted as an implementation of a Frank–Wolfe algorithm [68], taking the ℓ_1-distance as the nonsmooth concave merit function

$$\sum_{j \in I} \min\{x_j, 1 - x_j\}.$$

Based on this, in [52, 53] the use of other concave penalty functions for non-integrality was suggested. Therefore, the authors weighted the different terms of the distance function with coefficients that depend on the fractionality of the corresponding variable in the last LP solution.

In [35], a similar penalty system was used, and, additionally, the idea of performing several rounds of cutting plane generation to prevent the FP from cycling was introduced. This leads to fewer restarts and better and more solutions being found, but at the price of a significant increase in the complexity of the LPs (which are amended with cutting planes) being solved repeatedly, and hence in the total runtime. Note that in the original FP, each projection step consisted of solving one LP, which is now replaced by a series of LPs. When changing the objective (distance) function in the projection phase, one can use a warm-started primal simplex algorithm to solve the LP. When amending the projection step by cutting plane generation, the additional LPs should typically be solved by a warm-started dual simplex algorithm.

Finally, in [87] the authors proposed a FP variant in which a MIP search when the FP is about to cycle was introduced. The corresponding variable neighborhood pump can be interpreted as applying stage III of the FP variant [15] that is based on *local branching* [62]. The use of local branching to drive to feasibility partially feasible solutions obtained by FP was proposed in [63].

4.4 The Perturbation Step

The perturbation methods devised to avoid cycling are key ingredients of the FP method, which can greatly affect its behavior and success rate, both from the practical and theoretical points of view.

The original perturbation mechanism proposed in [61] works as follows. Whenever a cycle of length one is detected, that is, the integer point $\tilde{x}$ obtained at the current iteration is the same as the previous one, a (small) random number, say K, of binary variables is flipped to the opposite bound, so as to minimize the increase in distance between the two sequences. This is implemented by sorting the variables by nonincreasing fractionality in $\bar{x}$, and flipping the first K variables in $\tilde{x}$. The perturbation mechanism is designed to interfere as little as possible with the minimization of distance between the two sequences of points, basically exploiting the degeneracy in the distance function to pick a different integer point not too far away from $\tilde{x}$ (e.g., if all flipped variables had fractionality equal to 0.5, there would be no increase in distance at all).

If a longer cycle is detected, a different perturbation method, akin to a restart, is used, in which all binaries might be flipped according to some probability that depends on their fractionality.

The picture gets more complicated when general integer variables are present in the model. A careful analysis of the original source code in [15] reveals a quite elaborate restart scheme to decide how much a single variable has to be perturbed, and how many variables have to be changed. In particular, a single variable x_j is perturbed by taking into account the size of its domain: if $u_j - l_j < M$ with a suitable large coefficient M, then the new value is picked randomly from within the domain. Otherwise, the new value is picked uniformly in a large neighborhood around the lower or upper bound (if the old value is sufficiently close to one of them), or around the old value. The number of variables to be perturbed, say RP, is also very important and is changed dynamically according to the frequency of restarts, decreasing geometrically at every iteration in which restarts are not performed, and increasing linearly. Finally, RP is bounded by a small percentage (10%) of the number of general integer variables, and the variables to be changed are picked at random at every restart.

In addition, [15] also introduced a subtle random component in the rounding function. Instead of computing $\tilde{x}_j$ as $\lfloor \bar{x}_j + 0.5 \rfloor$, it is computed as $\lfloor \bar{x}_j + \tau \rfloor$, where τ is obtained as

$$
\tau(\omega) = \begin{cases} 2\omega(1 - \omega) & \text{if } \omega \le \frac{1}{2}, \\ 1 - 2\omega(1 - \omega) & \text{if } \omega > \frac{1}{2}, \end{cases}
$$

where ω is a uniform random variable in $[0, 1)$.

It is worth noting that the various extensions described in this chapter can have an influence on the need for a perturbation mechanism. In particular:

- in the objective FP there is no need to perturb the current integer point if we are in a phase in which α can still change;

- the different penalty functions described in [35, 52, 53], as well as the addition of cutting planes, can reduce the need for an explicit perturbation step;

- the rounding step based on constraint propagation described in [66] itself uses some randomization when ranking the fractional variables.

Convergence results for the method also heavily depend on the interplay between the projection, rounding and perturbation steps. The theoretical analysis in [54] showed that the method, when applied to mixed 0-1 programs, can fail to converge if the perturbation step is limited to flip only variables that are currently fractional. At the same time, flipping variables that are currently integers in a naïve way can also lead to non-convergence and poor results. By contrast, a perturbation method based on WALKSAT [156] was shown to yield a convergent method.

Without any perturbation, the method will still converge to a local minimizer of the distance between the two sequences (see, e.g., [35, 52, 73]); however this is not necessarily a feasible solution of the original model as the distance is usually strictly positive. The analysis in [35, 52, 73] is based on the interpretation of an *idealized* FP version, that is, without perturbation, in terms of classical algorithms from the literature. Namely, in [35] the idealized FP was interpreted as a discrete version of the *proximal point algorithm*, while in [52] as a *Frank–Wolfe algorithm* applied to a suitable chosen concave and non-smooth merit function and, finally, in [73] as an *alternating direction method*. In [73], such an interpretation allowed the authors to replace the perturbation by a penalty framework, thus preserving some form of convergence that is proper for alternating direction methods.

4.5 Objective Diving

A distinguishing feature of the FP approach is that it tries to enforce integrality not by fixing variables but rather by changing the objective function. The same approach can, in principle, be applied to the diving heuristics that we described in Chapter 3: the overall scheme is the same, but instead of *hard* fixing a variable to a value, we do a *soft* fixing instead, by changing the variable's objective value to a very large positive (resp. negative) value if we want to push the variable to its lower (resp. upper) bound. The method has advantages and disadvantages: on the positive side the resulting LP can never become infeasible and we can always *undo* a soft-fixing by changing the objective again; on the negative side the LPs do not get smaller (and thus faster to solve) along a dive, we can no longer apply constraint propagation, and we need some way to avoid cycling. Notice that LP warm-starting is unaffected: instead of using the dual simplex after a bound change, we can just use the primal simplex after an objective change.

Other objective diving heuristics are not very popular, with only a couple of variants appearing in the literature: *objective pseudocost diving*, the objective counterpart of pseudocost diving, and *root solution diving*, the objective counterpart of linesearch diving [1]. The current consensus is that they are not as effective as regular diving heuristics and they are somewhat dominated by the (more sophisticated) FP heuristics. Still, [1] reported good results for some classes of problems, so it might be worth some future research.

4.6 Quiz

It's quiz time! This section is designed to test your understanding of the concepts covered in this chapter. Each question is followed by four possible answers, only one of which is correct. If you are unsure about some questions, you might find it helpful to work through the corresponding section again. Remember, sometimes the process of elimination can be as valuable a tool as direct problem-solving. By engaging with this quiz, you'll reinforce your knowledge and identify areas requiring further study. Take your time, think carefully and check the answers in the Appendix when you are done.

Question 1 The FP generates two (hopefully convergent) sequences of points where

- ☐ both are LP feasible;
- ☐ both are integer feasible;
- ☐ one is LP feasible and one is integer feasible;
- ☐ one is feasible for one half of the constraints, and the other is feasible for the remaining half.

Question 2 What is the most crucial aspect when designing an FP-like algorithm?

- ☐ a quick rounding step;
- ☐ dealing with cycles;
- ☐ a quick projection step;
- ☐ the starting point.

Question 3 Which form of perturbation guarantees convergence of the original FP?

- ☐ any;
- ☐ flipping any subset of the fractional variables;
- ☐ a WalkSAT-like scheme, based on which constraints are actually violated;
- ☐ none.

5

Pivoting and Line Search Heuristics

This chapter discusses some primal heuristics that do not necessarily belong to the mainstream methods that have been implemented in the MIP solvers but are still quite interesting either for historical reasons – some of them represent the first attempts by the mathematical programming community to devise heuristic solutions within a general MIP scheme – or because they combine many of the ingredients that are extensively discussed in this book. In particular, we will discuss two families of algorithms: in Section 5.1, we consider the family of *pivoting heuristics* that originated from the seminal pivot and complement scheme [10]. In Section 5.2, we discuss a few examples of primal heuristics using the *line search* idea[1] from rather different angles.

5.1 Pivoting Heuristics

As mentioned, the area of pivoting primal heuristics was pioneered by the famous *pivot and complement* algorithm [10]. For a binary integer program, the basic observation is that each constraint

$$\sum_{j \in N} a_{ij} x_j \le b_i$$

can be rewritten as

$$\sum_{j \in N} a_{ij} x_j + s_i = b_i,$$

and if one is able to obtain through the bounded simplex method a solution where all artificial slack variables $s_i \ge 0$ are basic, then all original binary

---

[1] The line search idea has been already used in the context of diving heuristics; see Section 3.3.

69

variables are nonbasic at their bound, that is, they are integer and the solution is feasible. Then, given an initial solution $(\bar{x}, \bar{s})$ of the LP relaxation of the problem reformulated with the slack variables, the algorithm proceeds by *pivoting* in the basis slack variables while trying to maintain feasibility for the constraints. More precisely, three types of pivots are considered:

- A pivot of *type 1* is one that exchanges a basic x_j with a slack s_i and maintains primal feasibility. Of course, this is the ideal case.
- A pivot of *type 2* is one that – by maintaining primal feasibility – keeps the same number of basic variables x_j but reduces the overall *integer infeasibility* of the solution measured as

$$\sum_{j \in N} \min\{\bar{x}_j, 1 - \bar{x}_j\} \tag{5.1}$$

by a strictly positive quantity.[2]
- A pivot of *type 3* is one that sacrifices primal feasibility by exchanging a basic x_j with a slack s_i that enters the basis with a *negative* value.[3]

Because of the existence of some pivots of type 3, the initial solution $\bar{x}$ has possibly become infeasible, say $(\hat{x}, \hat{s})$. Then, the algorithm proceeds by *complementing* nonbasic variables, its second major ingredient, with the attempt of reducing the overall *primal infeasibility* measured as

$$\sum_{j \in N} \max\{0, -\hat{x}_j\} + \sum_{i=1}^{m} \max\{0, -\hat{s}_i\}, \tag{5.2}$$

which corresponds to the sum of the violations of the constraints. More precisely, a set of one or two nonbasic variables is complemented if (5.2) is reduced by a strictly positive quantity.

This procedure is called *search* phase and is concerned with finding a first feasible solution $\tilde{x}$. The algorithm includes an improvement phase that seeks to improving $\tilde{x}$. Overall, variable fixing based on reduced costs, rounding to the nearest 0-1 solution, and truncating operations (i.e., rounding any fractional value to 0) are also applied in the attempt to converge faster to feasible solutions or improve them (see [10] for details).

In [119], the idea of applying tabu search [77] principles to pivot and complement was explored. Essentially, type 3 pivots in the pivot and complement scheme are used to escape from local optima, that is, when no type 1 or type 2 options exist. Of course, this could potentially lead to cycles, and by using tabu search, escaping from local minima and cycling are managed by the tabu

---

[2] Recall that we are considering a pure binary program, thus $N = I = \{0, 1\}^n$.
[3] This pivot would be necessary if no pivots of type 1 or 2 exist.

list, so one is allowed to avoid type 3 pivots provided that some non-improving type 2 pivots are performed. In other words, one can maintain primal feasibility (guaranteed by type 1 and 2 pivots) at the price of not always reducing the integer infeasibility measure (5.1), and of course one needs to avoid cycling back to the previous solution, which is achieved by storing recent moves (pivots). Similar ideas are applied to the improving phase of pivot and complement.

More than a decade later, the pivot and complement idea was revisited and extended in [57]. The core component in [57] is still pivoting (where the sequence of pivots is interpreted as rounding the solution $\bar{x}$), but the integer infeasibility measure (5.1) is generalized to be a parametric *concave* function. The pivots are performed according to a few rules whose common idea is to privilege as much as possible pivots that decrease the concave integer-infeasibility function while limiting the increase of the real objective function $c^{\mathsf{T}}x$, ideally with no increase (rule 1). When that is not possible, a probing step is performed that takes into account the effect of all possible pivots in both the integer-infeasibility measure and objective function deterioration. When such a step also fails, then a cutting plane to remove the current solution $\hat{x}$ is added and the procedure is iterated. The approach is tailored to mixed binary programming problems and it is presented as a standalone heuristic, that is, no explicit way of embedding into a branch-and-bound algorithm is discussed.

Finally, in [75], Gomory cuts [79, 80] – see Section 1.2 – were used in conjunction with pivot and complement to guarantee a stronger diversification effect when type 1 and type 2 pivots fail. Essentially, the use of Gomory cuts allows us to avoid pivots of type 3 (leading to primal infeasibility) and complementation. More precisely, the idea is to solve the LP relaxation of the MIP (1.1), obtain $\bar{x}$, and at the same time separate a violated Gomory cut. Such a cut, say $\alpha^{\mathsf{T}}x \leqslant \beta$, is by definition not satisfied by $\bar{x}$, that is, $\alpha^{\mathsf{T}}\bar{x} > \beta$. Then, pivots similar to types 1 and 2 are performed with the caveat that, in the case that a pivot leads to a solution $\hat{x}$ that satisfies $\alpha^{\mathsf{T}}\hat{x} \leqslant \beta$, a new cut is separated. When no such pivots exist, then the most recent Gomory cut is added to the LP to restart the search.

5.2 Line Search Heuristics

The idea of line search algorithms is simple and, in the MIP context, can be summarized as follows. Given an optimal solution $\bar{x}$ of the LP relaxation, one draws a line connecting $\bar{x}$ to a second point, say, $\hat{x}$, chosen according to a specific criterion. Then, a sequence of, say, K discretized points in the line connecting $\bar{x}$ and $\hat{x}$ is sampled as $x^k := \bar{x} + \alpha_k(\hat{x} - \bar{x})$, with $k = 1, \ldots, K$

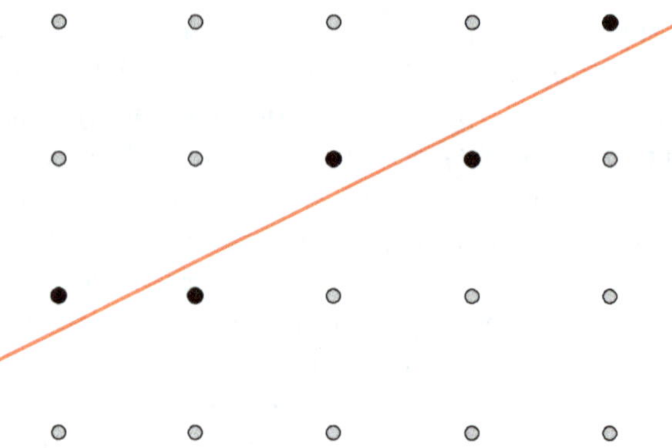

Figure 5.1 Generic line search. Integer points closer to the search line are marked in black.

and $0 = \alpha_1 < \cdots < \alpha_K = 1$. Each x^k of this discretized sequence is used to construct a solution candidate $\tilde{x}^k$. More precisely, a (mixed-) integer point $\tilde{x}^k$ is associated to each x^k, typically by rounding, and is then checked for feasibility. Local search operations on the variables of the candidate integer solutions $\tilde{x}^k$, $k = 1, \ldots, K$, are allowed in the attempt to improve the feasibility and/or quality of the candidate points. A pictorial representation of the general idea is given in Figure 5.1.

This scheme was originally introduced in the MIP context in [94] by selecting $\hat{x}$ in the attempt to go inside the polyhedron defined by the LP relaxation of P along a so-called *interior path*. Such a path is obtained by selecting $\hat{x}$ as the optimal solution of the LP relaxation of the problem in which the binding constraints[4] were modified by reducing the right-hand side vector b by an appropriate factor [94]. The underlying geometrical intuition, at least for general integer MIP instances, is that going inside such polyhedron increases the chances for the x^k points to be feasible.

In [100], the algorithm in [94] was extended, first of all, by considering the mixed-integer linear programming case versus the integer linear programming one, as in [94]. However, the major difference with respect to [94] is the way in which the line segment is defined. The idea is to solve the LP relaxation of an auxiliary linear program that is obtained from P by relaxing the integrality requirements and by (i) adding a unique variable λ used as a slack in all constraints, (ii) imposing the equality of the original objective function to a constant u, that is,

$$c^{\mathsf{T}}x = u, \tag{5.3}$$

and (iii) maximizing λ.

---

[4] Binding constraints are those of (1.1) satisfied at equality.

In the feasible region obtained by intersecting the LP feasible region with (5.3), the optimal solution of such an auxiliary problem[5] is the one that maximizes the minimum of the distances to the constraints of the original problem written in equality form. This geometric interpretation suggests that rounding this point (to the nearest integer plus some local search) has the potential to find feasible solutions. The first value of u is $u = c^\mathsf{T}\bar{x}$, and the line is defined by continuously increasing u.

In [58], the question on how to select $\hat{x}$ was approached from a statistical standpoint. In the original paper [94], the methods for defining $\hat{x}$ were obtained by considering the largest possible rounding distance either in terms of distance from $\bar{x}$ to its rounding or in terms of increase of the left-hand side of a binding constraint due to rounding. Instead, in [58] a less conservative approach is considered: by assuming that the parameters of the MIP instance (1.1) are drawn from some underlying probability distribution and that the fractionality of $\bar{x}$ are independent and identically distributed, one can define statistically defined perturbations of the vector b and the associated interior paths.

In [151], the line search idea was adopted with a slight variation. Technically, no extra point $\hat{x}$ is considered, but a direction toward the interior of the polyhedron defined by the LP relaxation is followed in the attempt to find a so-called 1-*ceiling point*. The concept of a ceiling point for a constraint[6]

$$\sum_{j \in N} a_{ij} x_j \le b_i$$

is that of an integer point $\tilde{x}$ that (i) satisfies the constraint for (ii) at least one direction h,

$$\sum_{j \in N} a_{ij} \tilde{x}_j + |a_{ih}| > b_i.$$

Essentially, condition (ii) says that by moving the integer point toward the constraint in (at least) one specific direction, the resulting integer point violates the constraint. Specifically, ceiling points can be characterized with respect to how many directions are simultaneously creating infeasibility, and the concept can be generalized to the entire set of constraints instead of one only. The algorithm in [151] starts from the solution of the LP relaxation $\bar{x}$ and finds one constraint to go toward[7] by following a direction that is obtained by a convex combination of the extreme directions (edges) tight in $\bar{x}$ as well as the constraint itself. Following the line given by such direction, steps are defined

---

[5] This auxiliary LP is parametric in the value of the constant u.
[6] This is in the spirit of variable locks; see Section 1.1.
[7] This is the reverse direction with respect to the way used to define the ceiling point.

as in regular line search, and rounding is used to obtain the 1-ceiling point for the considered constraint. Of course, being feasible for a specific constraint is not enough, and rounding is applied to ideally recover full feasibility. Finally, the algorithm can be extended to look for improving solutions once a fully feasible one has been found. The algorithm is designed to deal with models that do not include equality constraints.

In [9], the line search algorithm called *octane*, for <u>oc</u>tahedral <u>n</u>eighborhood <u>e</u>numeration, was introduced (again for binary IPs). The algorithm also starts with an LP optimal solution $\bar{x}$ and moves it along a certain search direction $\hat{x} - \bar{x}$. However, the integer (possibly infeasible) points discovered along this direction are not defined through rounding, but exploit the one-to-one correspondance between the vertices of the unit hypercube and the facets of the octahedron[8] that circumscribe it. More precisely, octane considers the center $x^0 = (\frac{1}{2}, \ldots, \frac{1}{2})$ of the unit hypercube and, for any of its vertices $\tilde{x}$, not necessarily belonging to the LP relaxation of P, it defines the hyperplane $H(\tilde{x})$ passing through $\tilde{x}$ and orthogonal to $\tilde{x} - x^0$. As anticipated, the half-spaces induced by these hyperplanes that contain x^0 define an octahedron.

Starting from $\bar{x}$ and going toward $\hat{x}$ and beyond, octane considers the first k hyperplanes $H(\tilde{x}^1), H(\tilde{x}^2), \ldots, H(\tilde{x}^k)$ that are intersected by the half-line $\hat{x} - \bar{x}$ and, exploiting the one-to-one correspondance above, checks for feasibility of the corresponding binary points $\tilde{x}^1, \tilde{x}^2, \ldots, \tilde{x}^k$. The best feasible solution $x^\star$ (if any) is updated accordingly.

The algorithm exploits an effective method for the enumeration of the first k facets of the octahedron and considers several options for the search direction $\hat{x}$, for example, the vector defined as the average of the set of extreme rays (normalized) of the cone defined by the current optimal basis of the LP relaxation and the inward normal to the objective function vector. Interestingly, when the algorithm is applied at any node of the branch-and-bound tree, the line can be defined as the difference between the optimal solution of the LP relaxation at the current node and $\bar{x}$ (in the spirit of line search diving; see Section 3.3). A pictorial representation of the heuristic is given in Figure 5.2.

Finally, in [139], the analytic center – see (1.8) in Section 1.6.1 – was used as $\hat{x}$ to define two line segments: one connecting $\hat{x}$ to $\bar{x}$ and the other connecting $\hat{x}$ to the LP relaxation solution obtained by inverting the objective function c. In the resulting algorithm, line search is performed on the two segments in the same spirit as that discussed in Section 4.2. If a feasible solution is not found,[9] the method changes the (weighted) analytic center by either changing

---

[8] Thus the name of the method.
[9] Or in the attempt of improving the best feasible solution $x^\star$.

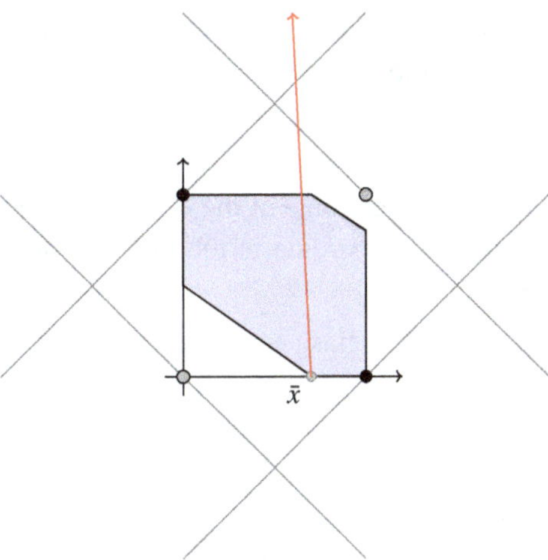

Figure 5.2 Octane heuristic. The shooting direction is marked in red.

the weights associated with the constraints or adding cutting planes with the aim of avoiding repeating the same rounding operations. With a new analytic center in place, the method is iterated.

5.3 Quiz

It's quiz time! This section is designed to test your understanding of the concepts covered in this chapter. Each question is followed by four possible answers, only one of which is correct. If you are unsure about some questions, you might find it helpful to work through the corresponding section again. Remember, sometimes the process of elimination can be as valuable a tool as direct problem-solving. By engaging with this quiz, you'll reinforce your knowledge and identify areas requiring further study. Take your time, think carefully and check the answers in the Appendix when you are done.

Question 1 Why is a pivot replacing an original basic variable with a slack beneficial in the pivot and complement heuristic?

- ☐ the slack variable becomes nonnegative;
- ☐ the original basic variable goes to one of its bounds;
- ☐ the original basic variable takes value zero;
- ☐ the objective function value improves.

Question 2 Can one apply line search starting from a feasible but nonoptimal solution of the LP relaxation?

- ☐ yes;
- ☐ no;
- ☐ yes, but only if the other extreme of the line is its rounding;
- ☐ yes, but only if the other extreme of the line is another feasible solution of the LP relaxation.

Question 3 The analytic center has been tested as one extreme point of a line search because

- ☐ it has favorable algebraic properties;
- ☐ it is easy to round;
- ☐ it is easy to compute;
- ☐ it has favorable geometric properties.

6

Computational Study

The present chapter constitutes the computational study of this book. We investigate how primal heuristics have an impact on the performance of a MIP solver, with respect to multiple performance measures.

Conducting an empirical study comes with a variety of choices that have to be made upfront:

- Which hardware to use?
- Which software to use?
- Which configuration of this chosen software to use?
- Which test instances to use?

Some of the choices are straightforward. Which hardware? The one available to the authors! Others are mostly based on community standards, for example, chosen time limits. Finally, some choices can have a big impact on the result and have to be taken and justified with care. The following paragraphs answer these questions one by one and thereby explain the setup for our experiments.

As a test set we used the benchmark set of MIPLIB 2017 [129], which consists of 240 publicly available instances. Since its first edition in 1992, MIPLIB has provided the MIP research community with a curated set of challenging, real-world instances from academic and industrial applications that are suitable for testing new algorithms and quantifying performance. It has become the gold standard in the community for conducting computational experiments with *general* MIP solvers.

At the time of writing this book, MIPLIB 2017 was the latest of, until then, six MIPLIB editions and therefore a natural choice for studying the impact of primal heuristics on a general MIP solver.

As for software choice, we settled for the publicly available MIP solver SCIP , which, as this book was being written, became completely open-source.

SCIP is currently, and has been for two decades, one of the fastest noncommercial MIP solvers; it is available in source code at www.scipopt.org/ and on GitHub. Its plugin-based framework design makes it easy and natural to isolate individual parts of the solution process, such as cutting, branching, or, in our case, primal heuristics. We deliberately decided on an open-source project for our experiments and against closed-source commercial solvers. The main reason is that SCIP allows direct control of each individual heuristic, whereas for commercial solvers, often only meta-controls exist. Also, with an open-source solver, one can check that all existing heuristics have been addressed by the set controls, while for closed-source products it is not always obvious what happens "under the hood."

That being said, all major solvers provide a control to deactivate all heuristics at the same time. This was used in a previous study on the impact of primal heuristics on MIP solver performance just over 10 years ago [17], with the commercial solvers FICO XPRESS, IBM ILOG CPLEX, GUROBI and the noncommercial solvers SCIP and CBC. While the final results of that particular experiment did not get published, the author reported that all solvers clearly benefited from heuristics, albeit to a different degree.

In our experiments, we will go one step further and rather than compare the effect of switching off all heuristics, we will study certain classes of heuristics. Next to running in default mode, we will run SCIP:

- without any sub-MIP based heuristics, that is, without LNS-based heuristics, such as in Chapter 2;
- without any heuristics that solve auxiliary LPs, that is, without diving and feasibility pump type heuristics, such as in Section 3.3 and Chapter 4; and
- without any LP-free heuristics, that is, without rounding and propagation-based heuristics, such as in Sections 3.1 and 3.2.

We decided against extending this study to individual heuristics, partially because of the sheer number (there are more than 30 primal heuristics implemented in SCIP), but most importantly because many heuristics in the same class have similar effects. If a single one is deactivated, others of the same class will take over its role, potentially at a slightly higher cost or being slightly less effective. However, the impact of deactivating a single heuristic is expected to be diminishing on a broad test set, by design. It is only when whole classes of heuristics are missing that we will be able to see their real impact.

For the computational results presented in this chapter, we used the latest SCIP available when this experiment was conducted. That was SCIP 8.0.2 [32], compiled with SoPLEX 6.0.2 [165, 174] as the LP solver, bliss 0.77 [102] for symmetry detection, and PaPILO 2.1.1 [76] as a presolving library. Thus, we exclusively used academic software available in source code for our experiments.

The results were obtained on a cluster of 64 bit Intel(R) Xeon(R) CPU E5-2670 v2 CPUs at 2.50 GHz with 12 MB cache and 128 GB main memory, running an Ubuntu 16.04.5 with a GCC 5.4.0 compiler. Turboboost was disabled. SCIP only ran single-threaded. Still, in all experiments, we ran only one job (SCIP instance) per node to reduce fluctuations in the measured runtimes that might be caused by interference between jobs that share resources, in particular the memory bus. For more details on hardware considerations for running MIP experiments (including considerations on *performance variability* [117]), we refer to the corresponding section in the two most recent MIPLIB papers [106, 129] or to the blog post [20].

For the computational experiments in this chapter, we used a time limit of two hours and a memory limit of 100 GB. The primal-dual (optimality) gap threshold was set to 0.0, which is actually the default for SCIP. Commercial solvers often use a small positive value, for example, 0.01%, as an optimality gap threshold for declaring an instance to be solved. For most applications, this is a reasonable approach, but it might be inappropriate in other situations. One problem with using a nonzero optimality gap threshold is that the results are not invariant under adding an offset to the objective function.

Taking all these considerations together, it is important to stress that these results are more a snapshot than set in stone. Different solvers, or even different versions of the same solver, would have given different data. Another test set, be it a past or a future MIPLIB release, or some completely different set of heterogeneous MIP instances, would probably have given different results. Nevertheless, we believe that the results presented in this chapter are well suited to reach an understanding of the performance impact of primal heuristics and individual classes of primal heuristics, while at the same time, it is impossible to reduce these insights to a single, precise number.

6.1 The Impact of Primal Heuristics

In SCIP, each heuristic has a frequency parameter called `freq` that determines after how many nodes a heuristic will be called again after its last call. Very cheap heuristics typically have a frequency of 1, that is, they are called at every node. Diving heuristics often have a frequency of 10 or 20, which means they will be called at every tenth or twentieth node, respectively. Some expensive start heuristics have a frequency of 0, which means that they will be called a single time only, typically at the root node. A heuristic is disabled by setting its frequency to -1.

Table 6.1 *LP-free heuristics in* SCIP.

intshifting	oneopt	randrounding	rounding	shiftandpropagate
shifting	simplerounding	trivial	zirounding	

Table 6.2 *LP-based primal heuristics in* SCIP.

adaptivediving	conflictdiving	distributiondiving	farkasdiving
feaspump	fracdiving	guideddiving	linesearchdiving
objpscostdiving	pscostdiving	rootsoldiving	veclendiving

Further, SCIP classifies all of its heuristics by display characters as either "d"iving, "o"bjective diving, "p"ropagation, "r"ounding, "t"rivial, "i"terative or "l"arge neighborhood search heuristics. We compiled our parameter settings following these classifications and set all heuristic frequencies of certain classes to -1 for disabling the respective heuristic classes.

We opted for separating the heuristics into the classes of LP-free heuristics, LP-based heuristics and MIP-based heuristics.

We consider a heuristic LP-free if it does not solve any auxiliary linear program, with the possible exception of one final LP to find a value for the continuous variables of a MIP once all other variables have been assigned a value. This comprises rounding heuristics as described in Section 3.1, propagation heuristics as in Section 3.2, iterative improvement heuristics like the 1-opt and 2-opt heuristics mentioned in Section 1.5.2, and a trivial start heuristic that tries some standard solutions such as the all-zero solution or all variables set to their upper bound.

The list of LP-free heuristics in SCIP includes the nine heuristics in Table 6.1. Note that this and the lists in Tables 6.2 and 6.3 only contain heuristics that are active for MIP solving by default and do not contain heuristics that are only used for other classes of optimization problems, for example, nonlinear problems.

LP-based heuristics are allowed to solve auxiliary LPs and typically solve a sequence of linear programs as the main ingredient of their logic. Typically, these heuristics solve a variation of the orignal LP relaxation that has been amended by a few variable fixings or an objective function change. They base the decisions to be taken in one iteration on the solution of the LP solved in the previous iteration. The diving heuristics described in Section 3.3 fall into this category, as well as the feasibility pump heuristics from Chapter 4 and other objective diving heuristics.

The list of LP-based heuristics in SCIP includes the 12 heuristics in Table 6.2.

Table 6.3 *MIP-based primal heuristics in* SCIP.

ALNS	Crossover	LP-face
GINS	RENS	RINS

Table 6.4 *Impact of primal heuristics (*Miplib*2017 benchmark).*

	all					all feas		all opt	
	feas	opt	obj	t_{heur}	$\frac{P(t_{max})}{t_{max}}$	$time_1$	sols	nodes	time
			(235 instances)			(175 instances)		(102 instances)	
default	213	124	–	11.6%	18.7%	14.0	36.6	2 060	164.9
alloff	184	116	7:51	–	35.1%	83.7	11.9	2 955	195.9
nolp	206	122	8:37	2.2%	25.2%	19.9	27.5	2 098	162.7
nolpfree	213	122	21:18	9.4%	22.3%	22.5	24.7	2 203	177.4
nomip	212	119	4:26	10.5%	20.7%	14.3	35.7	2 298	175.1

All MIP-based heuristics solve at least one auxiliary MIP, sometimes as small series of MIPS. These typically stem from fixing many variables, adding very restrictive constraints and changing the objective function such that it leads to follow-up fixings. The large neighborhood search heuristics described in Chapter 2 fall into this category.

The list of MIP-based heuristics in SCIP includes the six heuristics in Table 6.3. For our experiment of deactivating MIP-based heuristics, we furthermore disabled the solution of a final sub-MIP in the clique, locks and vbounds heuristics, so that they will act as pure propagation heuristics, that is, LP-free heuristics.

The results of the computational experiments are summarized and aggregated in Table 6.4. Each line of the table represents a different setting. Row "default" shows the standard behavior of SCIP with all heuristics running that are activated by default and no user settings applied. Row "alloff," to the contrary, summarizes the results of SCIP running without any primal heuristics; they have all been deactivated by setting their frequency parameters to -1. The next three lines aggregate SCIP's results when certain groups of heuristics are deactivated: "nolp" for deactivating all LP-based heuristics but keeping the other two types at their default, "nolpfree" for deactivating all LP-free heuristics while having others at default, and "nomip" for deactivating MIP-based heuristics plus applying the aforementioned settings to the clique, locks and vbounds heuristics, while everything else is left at default.

To aggregate numbers that spread among different orders of magnitude, such as the runtime, we use the so-called shifted geometric mean.

Definition 6.1 (shifted geometric mean)   Let $n \in \mathbb{Z}_{\geqslant 0}$, $\mathcal{V} = \{v_1, \ldots, v_n \mid v_i \in \mathbb{R}_{\geqslant 0}$ for all $i\}$ and $s \in \mathbb{R}_{\geqslant 0}$. The *shifted geometric mean* of $\mathcal{V}$ with shift s is defined as

$$\phi(\mathcal{V}, s) := \sqrt[n]{\prod_{i=1}^{n}(v_i + s)} - s.$$

Unlike arithmetic means, a geometric mean prevents large numbers, for example, hard instances at or close to the time limit, from having a huge impact on the aggregated measure. Further, using a shift reduces the effect of very small numbers, for example, for times, we use a shift of $s = 10$ (seconds) to reduce the impact of very easy instances in the mean values. Thus, the shifted geometric mean has the advantage that it reduces the influence of outliers in both directions. As a rule of thumb, the shift is often chosen one to two orders of magnitude smaller than the geometric mean of the numbers that are taken into account. Consequently, we used a shift of 100 for the number of branch-and-bound nodes, 10 for all times, and 1 for the number of solutions and the heuristic time ratio "t_{heur}."

In [18], the important argument was made that one should not concentrate on a single performance measure to evaluate the impact of primal heuristics on MIP solvers, but instead several criteria should be taken into account simultaneously. Consequently, we depict nine different statistics for each configuration, arguably not all of them being performance metrics.

In Table 6.4, columns "feas" and "opt" state for how many instances a feasible solution was found and how many instances were solved to proven optimality within the time limit, respectively. Bigger numbers are better. Column "obj" depicts for how many instances a setting produced a primal solution that was either at least 10% better than the best solution found with the default settings or at least 10% worse. A ratio where the first number is larger than the second would be favorable for the setting; if the second number is larger, default wins in this measure. Column "t_{heur}" shows which percentage of runtime was spent for primal heuristics in shifted geometric mean. This is not really a performance measure but rather indicates the trade-off that the MIP solver takes. How much of its runtime is invested in primal heuristics rather than in the branch-and-bound search or other procedures? Column "$\frac{P(t_{max})}{t_{max}}$" gives the average primal gap, as defined by the primal integral; see Definition 6.1. A smaller number is better here. Note that these first five columns refer to only 235 out of the 240 MIPLIB2017 benchmark instances. We excluded five instances for which one of the tested non-default settings failed.

Two statistics are given for the set of 175 instances for which *all* settings found a feasible solution (shown in the double-column "all feas"): "time$_1$," the mean runtime for finding a first feasible solution, and "sols," the mean number of primal solutions per instance that were found with a certain setting. Finally, for all 102 instances that were solved to optimality by *all* settings ("all opt"), we give the shifted geometric means of the number of branch-and-bound "nodes" and the overall running "time" to proven optimality.

From Table 6.4, we see that for all four non-default settings, almost all performance indicators speak in favor of using the respective primal heuristics that have been deactivated in those runs.

Let us first consider the three measures that simply "count" instances: "feas," "opt," and "obj," that is, for how many instances does a setting find a feasible solution, an optimal solution, or a better/worse solution than default, respectively. The default settings wins against the four non-default settings in all three categories, with only one exception: the setting without LP-free heuristics is slightly more likely to terminate with a better solution than the default.

When switching off individual groups of primal heuristics, the number of instances for which a feasible solution could be found within the time limit does not change a lot; it varies between 206 and 213, where the latter corresponds to the default. Only if we switch off *all* heuristics does this number drop significantly, namely to 184.

The number of instances solved to proven optimality is also hardly affected. It is 124 by default, varies between 119 and 122 for those settings where individual heuristic groups have been deactivated, and is lowest, at 116, for the setting without any heuristics. The statistics of the "obj" column complement the "opt" numbers given that they consider those instances that could not be solved to proven optimality and compare the incumbent solution quality when reaching the time limit. The "obj" comparison shows a clear benefit of LP-based and MIP-based heuristics – when deactivating them, there are significantly more cases where the primal bound at termination is inferior to the default setting than there are cases where a better bound is found. For LP-free heuristics, however, the default loses by a small margin. This can be partially explained by the fact that LP-free heuristics are mainly responsible for finding first, often quite bad, solutions in the beginning of the search, and thus a big impact on the final solution is not to be expected.

These three measures emphasize that primal heuristics mainly have an impact on the primal side of the problem. In our results, the number of instances that can be solved within the time limit hardly depends on primal heuristics, though for those instances that cannot be solved, primal heuristics very often lead to better solutions at termination. This can be seen as an indicator that applying primal heuristics is particularly worthwhile for hard instances that are not expected to be solved to proven optimality within a reasonable amount of time, which is a typical situation for real-world applications.

Besides the primal bound at termination, the time to find a first solution is the performance measure with the largest differences. When disabling primal heuristics, the time to prove feasibility increases by a factor of six ($83.7/14.0$). The effect of individual heuristic groups is still big for LP-based and LP-free heuristics, but is less pronounced. We can clearly recognize the effect that different heuristics compensate for each other: if one group is missing, the other group takes over the role of finding a first solution, and vice versa. It is only without any heuristics that finding a first solution becomes a challenge. Finally, MIP-based heuristics do not play a big role, which is not a surprise given that almost all MIP heuristics are improvement heuristics, with the notable exception of RENS.

Unsurprisingly, the number of feasible solutions found tells a similar story. When deactivating all heuristics, the number of solutions found drops by a factor of three ($36.6/11.9$). When only deactivating a single group of heuristics, however, there are still enough other heuristics present that generate solutions. The least impact again can be seen when deactivating MIP-based heuristics. These are typically called least frequently and hence provide the least solutions (but those are often of high quality, as shown by comparing columns "obj" and "opt").

Figure 6.1 visualizes the results from Table 6.4 by presenting the relative difference of the four non-default settings against the default setting with respect to six of the seven performance measures. We computed the relative difference in such a way that a positive value always corresponds to a deterioration:

- For measures for which larger numbers are better ("feas," "opt," "sols"), we took the value of the default setting minus the value of the non-default setting and divided the difference by the value of the non-default setting.
- For measures for which smaller numbers are better ("$P(t_{max})/t_{max}$," "$time_1$," "time," "nodes"), we took the value of the non-default setting minus the value of the default setting and divided the difference by the value of the default setting.

Figure 6.1 uses a logarithmic scale. Thus, we omitted relative differences below one percent from the presentation. For the same reason, the only degradation (-1.3% time to optimality for the "nolpfree" setting) is not shown in the diagram. Figure 6.1 captures vastly different orders of magnitude for the different performance measures and settings. The smallest is only 1.6% for

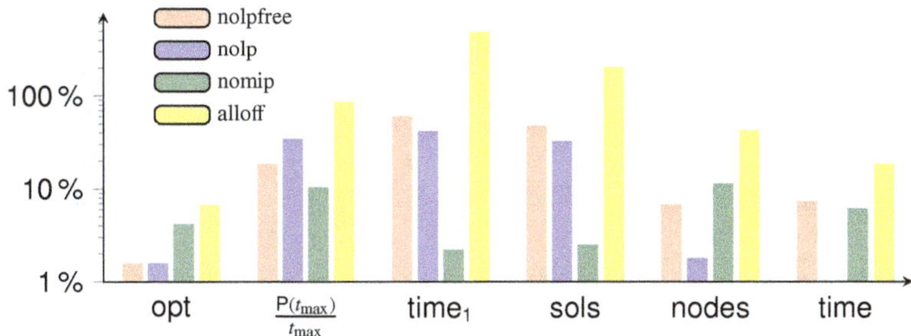

Figure 6.1 Bar chart illustrating the relative difference between SCIP running with and without primal heuristics with respect to six performance measures scaled such that positive values correspond to deteriorations by *not* using a class heuristic (or any).

the number of optimal solution with the "nolp" and "nolpfree" settings. The largest is a factor of six for time to find a first solution with the "alloff" setting. Figure 6.1 illustrates that for all four non-default settings, performance improves for at least six out of seven measures. We conclude that all types of primal heuristics are important in their own way.

There are performance measures such as the primal integral that show a heavy impact from all types of heuristics. For others, the picture differs a lot between different types of heuristics: while "$time_1$" is much less affected by MIP-based heuristics, those have the largest impact on the number of nodes. Our study confirms the 10-year-old results of [17] and [18] that it takes multiple measures to capture the real value of primal heuristics and that the two most common ones, the number of solved instances and average runtime, are not the most suitable ones.

For studying the primal integral in more detail, Figure 6.2 shows the evolution of the primal gap as a function over time. The blue dotted line corresponds to the average primal gap function, when running SCIP in default mode, with primal heuristics activated. The area under this curve corresponds to the average primal integral of this setting. The red dashed line shows the average primal gap function of the setting of SCIP with all heuristics deactivated, and the black, yellow and green lines correspond to SCIP running without LP-free, without LP-based and without MIP-based heuristics, respectively.

The results in Figure 6.2 clearly show that the primal integral is dramatically affected by switching off heuristics altogether. The other three settings that lack only one class of heuristics are closer to the SCIP default but are still clearly inferior. Figure 6.2 confirms that this picture is consistent over the whole one

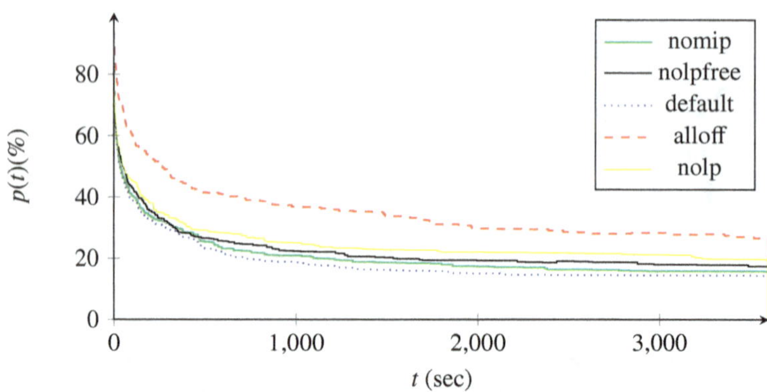

Figure 6.2 Evolution of the primal gap over time when running SCIP with and without primal heuristics on MIPLIB2017.

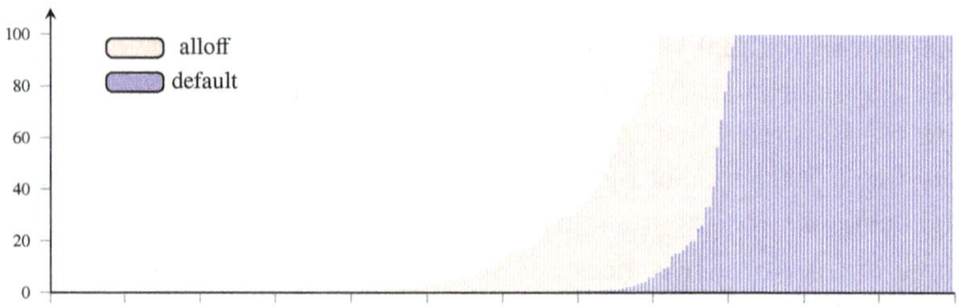

Figure 6.3 Percent of nodes to first solution.

hour of potential runtime considered in our experiment. Choosing a smaller time limit would not have changed the conclusion.

Finally, Figure 6.3 visualizes the impact of primal heuristics on the number of nodes needed to find a first feasible solution. The chart shows one bar for each instance and each of the two settings. The height of a bar indicates after how many percent of the nodes the first feasible solution was found. Red and blue bars show the performance without ("alloff") and with (SCIP default) primal heuristics, respectively. For each setting, the instances have been sorted by the measured percentage in non-decreasing order. Therefore, a red and a blue bar at the same position do not necessarily correspond to the same instance.

We observe that in Figure 6.3 the blue line of "default" bars clearly lies to the right of the red "alloff" bars. Having a closer look at the blue bars, we observe a rather pronounced "L-shape," or rather, a ⌐-shape: for most of the instances, the first solution is either found very early (mostly at the root node) or not until the very end. Note here, that the set of instances for which the ratio is 100%

compiles many different cases: pure feasibility instances, infeasible instances, instances that get solved at the root, and instances for which no solution is found before the time limit.

That instances either fall in a "trivial" or into an "unsolvable" (within reasonable time) category is a typical situation for $\mathcal{NP}$-hard problems, and it is part of the art of test set compilation to find a good middle ground. Note that MIPLIB2017 is not necessarily designed to be the best test set for finding first feasible solutions, but rather for solving instances to proven optimality.

6.2 Experiments with Additional Configurations

In this section, we consider a couple of more specific configurations that are useful to evaluate the effect of primal heuristics. More specifically, in Section 6.2.1, we run SCIP by activating a meta-parameter that gives emphasis to primal feasibility, and we compare such a configuration with the default and with a SCIP version in which heuristics are executed aggressively. A meta-parameter is a way to set a curated list of parameter settings all at once. When changing a meta-parameter, the solver often changes a few dozen individual parameters under the hood. In Section 6.2.2, we consider deactivating either the heuristics in charge of finding the first feasible solution or those that are used to improve the incumbent; hence, we compare the impact of the start against the impact of improvement heuristics.

Tables 6.5 and 6.6 in this section are organized the same way as Table 6.4; for a detailed explanation, see Section 6.1.

6.2.1 Feasibility Emphasis and Aggressive Mode

First of all, we would like to note that an emphasis on feasibility and aggressive heuristics are not necessarily the same. Aggressive heuristics would apply all heuristics more aggressively; for a feasibility emphasis, a particular focus might be on start heuristics. An emphasis on feasibility also *de-emphasizes* methods that primarily aim at improving the dual bound, for example, cutting plane generation or strong branching. Aggressive heuristics will only affect heuristic parameters. Such emphasis settings are not a peculiarity of SCIP but quite common to all state-of-the-art solvers.

Before analyzing the results in more detail, it is important to point out one thing: these meta-settings are non-default for a reason. If an emphasis on feasibility or aggressive heuristics would improve performance in all aspects, those would certainly be the default in the first place. At the same time, they exist for

Table 6.5 *Impact of primal heuristics (*MIPLIB*2017 benchmark), with different emphasis settings.*

	all					all feas		all opt	
	feas	opt	obj	t_{heur}	$\frac{P(t_{\max})}{t_{\max}}$	time_1	sols	nodes	time
			(237 instances)			(204 instances)		(102 instances)	
default	213	124	–	11.4%	18.7%	15.0	32.4	2 054	188.4
emphfeas	214	112	19:30	15.3%	19.9%	8.9	50.7	3 217	238.5
heuraggr	215	118	18:21	18.6%	20.5%	10.4	37.7	1 911	227.7

a reason. Certainly (like almost every non-default setting), they are good for *some* instances. The question that we ask ourselves here is whether there are measures for which they are good on average. If yes, which are those measures, and which are the ones that suffer?

Consider Table 6.5. A first thing to notice is that "opt" and "time," hence the columns showing the number of instances solved to optimality and the average time to optimality, clearly deteriorate for both meta-settings, as expected. Unsurprisingly, the time spent in heuristics increases in either setting and both settings produce more feasible solution than the default. There are performance measures in which the meta-settings beat the default: one is "time_1," the time to find a first solution, and the other is the related "feas," the number of instances for which a feasible solution was found.

It is interesting to observe how the two arguably related settings differ from each other. For example, the feasibility emphasis setting takes more than 50% more nodes to solve as compared to the aggressive heuristics setting. This is perfectly explained by the fact that the feasibility emphasis will deactivate or de-emphasize several dual techniques.

It is somewhat surprising that the primal integral does not benefit from SCIP's emphasis settings. This suggests that those meta-controls might benefit from fine-tuning in this direction. For the commercial solver FICO XPRESS, using the heuremphasis mode improves the primal-dual integral by about 30% (while deteriorating the time to optimality by about the same order).

6.2.2 Comparing Start and Improvement Heuristics

As a final experiment, we will look at the impact of deactivating all start versus all improvement heuristics. In this context, we call a heuristic an improvement heuristic when it requires at least one feasible MIP solution as an input, for example, RINS. A start heuristic is a heuristic that starts from scratch, not requiring a feasible MIP solution. Note that, in this notation, a heuristic is

Table 6.6 *Impact of primal heuristics (MiPLIB2017 benchmark), comparing improvement against start heuristics.*

	all					all feas		all opt	
	feas	opt	obj	t_{heur}	$\frac{P(t_{max})}{t_{max}}$	$time_1$	sols	nodes	time
			(234 instances)			(176 instances)		(106 instances)	
default	213	124	–	11.6%	18.7%	15.0	36.0	2056	176.3
noimpro	213	124	8:18	11.2%	19.4%	14.8	35.5	2135	183.1
nostart	184	112	8:45	1.1%	33.9%	86.4	10.4	2941	194.2

still considered a start heuristic, even if it is called in a situation where an incumbent already exists. The point is that a start heuristic does not make use of that incumbent.

Table 6.6 *seems* to indicate that start heuristics are much more influential than improvement heuristics. Before going into these results in more detail, we should note that these two experiments are not completely independent. Start heuristics work very well without improvement heuristics; improvement heuristics, however, will work significantly worse when there are no start heuristics to provide decent reference points to improve upon.

Consequently, we see that the number of problems for which a feasible solution is found does not change when switching off improvement heuristics, but it reduces drastically when deactivating start heuristics. Similarly, the time to find a first feasible solution increases by a factor of almost six without start heuristics. Note that deactivating improvement heuristics is not an operation without effects until the first feasible solution: for example, the shared resource limits in ALNS will be distributed differently and the exploration-exploitation tradeoff will be different for a changed set of heuristics.

The previous observations go along with the lack of start heuristics having a much bigger impact on the primal integral: the primal gap starts closing much later without them and, of course, the biggest jump in primal gap is often from a 100% gap to a first feasible solution. Later improvements might have a less pronounced impact on that measure.

The experiment still shows a clear benefit of improvement heuristics. The primal integral, number of nodes and time to prove optimality all deteriorate by about 4% without them. There are more than twice as many instances that terminate with a worse solution than with a better solution without improvement heuristics.

One final interesting observation is how imbalanced the time spent in improvement versus start heuristics is in SCIP. Arguably, it makes sense that more time is spent in the type of heuristics that have a bigger impact if they fail: start heuristics. Also, there are more of them and they are called much more frequently than most improvement heuristics. Again, these experiments are not independent: when deactivating start heuristics, improvement heuristics will start to be active much later in the search process. When comparing "$time_1$" and "time" in the default setting, we can roughly estimate (acknowledging that these are not taken on the same instance sets), that, by default, improvement heuristics could be called during more than 90% of the solution process, namely after the first solution was found. Without start heuristics, this reduces to a bit more than a half of the solution process.

Also, there often is a significant difference between the solutions that the tree search finds and those that start heuristics find. The latter are often of inferior quality and rely on improvement heuristics to improve them. This has a trickle-down effect on improvement heuristics: when the reference point is of high quality already, the sub-MIPs of LNS heuristics, for example, are more likely to become infeasible and not use a lot of time. Also, the frequency of calling improvement heuristics often depends on previous successes. If they are mostly called with close-to-optimal solutions, the success rate likely decreases.

After all, it can be said that the heuristics, like other parts of the solver, are well fine-tuned against each other and it is often not easy to completely single out an effect.

6.3 Quiz

It's quiz time! This section is designed to test your understanding of the concepts covered in this chapter. Each question is followed by four possible answers, only one of which is correct. If you are unsure about some questions, you might find it helpful to work through the corresponding section again. Remember, sometimes the process of elimination can be as valuable a tool as direct problem-solving. By engaging with this quiz, you'll reinforce your knowledge and identify areas requiring further study. Take your time, think carefully and check the answers in the Appendix when you are done.

Question 1 The main impact of primal heuristics can be seen in

- ☐ improving time to optimality;
- ☐ improving the dual bound;
- ☐ improving the primal bound early in the search;
- ☐ proving infeasibility.

Question 2 To balance the impact of improvements of easy and hard instances when computing means one can use

- ☐ a geometric mean;
- ☐ an arithmetic mean;
- ☐ a shifted geometric mean;
- ☐ a shifted arithmetic mean.

Question 3 By far the biggest impact on various success measures can be seen by

- ☐ LP-free heuristics;
- ☐ LP-based heuristics;
- ☐ MIP-based heuristics;
- ☐ the mix of all types of heuristics.

Question 4 Without primal heuristics, the primal integral of SCIP 8 on MIPLIB 2017

- ☐ hardly changes;
- ☐ increases by about 10%;
- ☐ almost doubles;
- ☐ increases by more than a factor of 10.

7

Primal Heuristics for Mixed-Integer Nonlinear Programming

As we saw in Chapter 6, primal heuristics play a crucial role in solving MIPs. For mixed-integer nonlinear programs (MINLPs), they are even more pivotal since having tight bounds early in the solution process is often essential for constructing tight relaxations. While many MINLP heuristics are extensions of those developed for MIPs, there also exist heuristics that are unique to MINLP and specifically designed to handle the complexities of nonlinear optimization.

Definition 7.1 (MINLP) A *mixed-integer nonlinear program (MINLP)* is an optimization problem of the form

$$
\begin{aligned}
\min \quad & f(x) \\
\text{s.t.} \quad & g_i(x) \leqslant 0 && \text{for all } i \in \mathcal{M}, \\
& l_j \leqslant x_j \leqslant u_j && \text{for all } j \in \mathcal{N}, \\
& x_j \in \mathbb{Z} && \text{for all } j \in \mathcal{I},
\end{aligned}
\tag{7.1}
$$

where $\mathcal{I} \subseteq \mathcal{N} := \{1, \ldots, n\}$ is the index set of the integer variables, $f \colon \mathbb{R}^n \to \mathbb{R}$, $g \colon \mathbb{R}^n \to \mathbb{R}^m$, and $l \in (\mathbb{R} \cup \{-\infty\})^n$, $u \in (\mathbb{R} \cup \{+\infty\})^n$ are lower and upper bounds on the variables, respectively.

Class (7.1) can be seen as the most general class of optimization problems. The special case that $\mathcal{I} = \varnothing$ is referred to as nonlinear programming (NLP). MIP is the special case of (7.1), where f and $g_i(x)$ are linear functions for each $i \in \mathcal{M}$. When this is not the case, we do have a genuine MINLP problem and its difficulty is highly related to the convexity of its continuous relaxation. More precisely, when $g_i(x) \leqslant 0$ defines a convex set for each $i \in \mathcal{M}$ and $f(x)$ is a convex function on the domain of x, problem (7.1) is referred to as *convex* MINLP, and as *nonconvex* otherwise. It is easy to see that nonconvexities complicate both the theoretical and practical solvability of optimization problem. It is $\mathcal{NP}$-hard, actually even undecidable, to optimize a nonconvex function even without constraints or to determine the feasibility of a single nonconvex constraint (without objective). Taking the complexity aside (MIP is also $\mathcal{NP}$-complete, after all), the size of nonlinear optimization problems that state-of-the-art global solvers like SCIP or FICO XPRESS can reliably solve is at least an order of magnitude smaller than for linear MIPs.

In the context of solving nonconvex nonlinear optimization problems, a solver is often explicitly referred to as a *global solver*, if, loosely speaking, given infinite time and other resources, the implemented algorithm will find a (proven) global optimum to any given optimization problem. This distinguishes global solvers from *local solvers* that may terminate with a solution that only fulfills some form of local optimality guarantee (typically requiring that certain regularity conditions are satisfied). Since for convex constraints and objective functions local optimality implies global optimality, such a differentiation is not common for convex or even linear programming.

Several of the papers cited in this chapter formulate MINLP with having a linear objective. Note that the above notation can be easily transferred into this alternative format by moving a nonlinear objective function into the constraint set and replacing it by a single, artificial variable: $\min x_{\mathrm{aux}}$, such that $f(x) \leqslant x_{\mathrm{aux}}$.

For nonlinear problems, having tight bounds, both on the objective and on individual variables, is even more crucial than for MIPs, because algorithms to solve MINLPs often rely on linear approximations or relaxations of nonlinear functions. The stronger such relaxations or approximations within a variable domain are, the tighter the domain. A better primal objective helps to tighten variable domains, for example, by reduced cost fixing [140].

Consequently, there has been important research activity on the topic of primal heuristics for MINLP, which we will review in the remainder of this chapter. Given both the complexity of the problem class and the fact that it often requires a significant amount of tree search to solve even small MINLPs, the research focus has been more on rather heavy heuristic strategies such as feasibility pumps and large neighborhood search heuristics and less so on quick and fast strategies such as fix-and-propagate or rounding. In Section 7.1, we present several adaptations of the feasibility pump framework, while, in Section 7.2, we discuss the large class of LNS-based primal heuristics. Finally, Section 7.3 presents a heuristic called undercover.

7.1 Feasibility Pumps

In Chapter 4, we described various variants of feasibility pump (FP) algorithms for MIP. We observed that the core concept behind all feasibility pump algorithms involves creating two sequences of points that aim to converge toward a feasible solution for a given optimization problem. One sequence includes points that satisfy the constraints for a continuous relaxation (the LP relaxation in the MIP case), yet may not satisfy integrality. The other sequence contains points that are integral but could violate some constraints of the original optimization problem. The subsequent point in one sequence is always derived by minimizing its distance from the last point in the alternate sequence. As a result of this, the used distance function may be different in each of the two main steps, which we also refer to as rounding and projection.

When designing an FP for MINLP, the main task is to adapt the two FP steps for nonlinearity. Two natural questions to ask are: (i) what kind of relaxation should be solved in the projection step, and (ii) is there a different way to perform the rounding step? *Nonconvex* MINLPs represent an extra burden: even the continuous relaxation might be disconnected and is, in general, nonconvex; hence, finding an optimum over such a relaxation might not be tractable.

The first two MINLP versions of the FP were presented in [37] and [38]. Both teams of authors considered *convex* MINLPs and implemented their ideas using Bonmin [36].

The paper [38] is probably the closest to the original FP. It keeps the rounding step as in [61] and replaces solving an LP in the projection phase by solving a convex NLP, using the original distance function (4.2) as an objective. The perturbation scheme is less aggressive than the one of [61], flipping only a single variable.

In [37], the authors suggested using an ℓ_2-norm for the projection step. Further, their implementation of the rounding step differs significantly from all previous FP variants. Instead of performing an instant rounding to the nearest integer, they solve a MIP relaxation that is based on an outer approximation [56] of the underlying MINLP,

$$\tilde{x} = \operatorname{argmin}\{\Delta(x, \bar{x}) \mid g(\bar{x}) + J_g(\bar{x})(x - \bar{x}) \leqslant 0, x \in [l, u], x_j \in \mathbb{Z}\ \forall j \in I\}, \quad (7.2)$$

where $J_g(\bar{x})$ denotes the Jacobian of the constraint functions (summarized to a single function $g \colon \mathbb{R}^n \mapsto \mathbb{R}^m$) evaluated at the NLP optimum $\bar{x}$. Solving a MIP relaxation, despite being an $\mathcal{NP}$-hard problem itself, is often computationally much cheaper than solving the original MINLP (compare, e.g., [130]). Interestingly, the two norms have switched roles in this FP version: where in [61] and [38], the ℓ_1-norm was used for the projection step and the ℓ_2-norm was

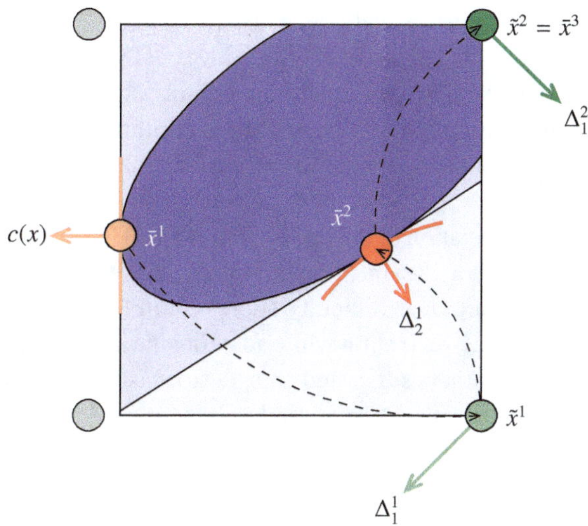

Figure 7.1 Feasibility pump for MINLP: the truncated blue ellipsis shows the feasible region of the NLP relaxation, the light blue trapezoid depicts the MIP relaxation for the second iteration, the very light blue rectangle (which includes the trapezoid) is the MIP relaxation after the first iteration. The sequence of NLP optima is shown as red points, the sequence of MIP optima as green points. Each red and green point also indicates for which function it is optimal: the original objective ($c(x)$, light red), an ℓ_2-distance function (Δ_2^1, dark red and curved) and the ℓ_1-distance functions (Δ_1^1 and Δ_1^2, green). Used with permission of Elsevier Science & Technology Journals, from [28]; permission conveyed through Copyright Clearance Center, Inc.

used for rounding, the opposite holds for [37]. An illustration of the algorithm is given in Figure 7.1. Note that, for this FP, both steps use a distance function Δ. We denote the Manhattan distance used for the rounding step by Δ_1 and the Euclidean distance of the projection step by Δ_2, using superscripts for the iteration count.

Solving (7.2) instead of performing a simple rounding $\tilde{x} = [\bar{x}]$ of course gives rise to better integral points (since feasibility is explicitly addressed in the rounding step), and has an important effect with respect to the main weakness of FP algorithms: cycling. For convex MINLPs, it is always possible to derive a cut,

$$(\bar{x} - \tilde{x})^\mathsf{T}(x - \bar{x}) \geq 0,$$

and add it to the MIP (7.2). By this, cycling is avoided. However, in the computational results presented in [37], the FP did not cycle even without these cuts. Because of the cycling avoidance, this version of the FP either finishes with a feasible point or proves that none exists.

To improve the solution quality of FPs for convex MINLP, [159, 160] adapted the idea of the objective FP [5] and used a combination of the original objective and an ℓ_1-distance function to extend the algorithm of [38]. The authors also experimented with using a local linearization of the original objective at the point computed in the previous FP iteration.

Similar to the interpretation of the linear FP as a Frank–Wolfe algorithm [52, 53], the paper [50] gave a classification of a nonlinear FP as a successive projection method. The particular difficulty addressed in [49, 50] is that of handling the nonconvex NLP relaxation while adapting the algorithm of [37] to the nonconvex case. The authors suggested using a stochastic multistart approach, feeding the NLP solver with different randomly generated starting points and solving the NLP to local optimality as if it was a convex problem. In the event that this does not lead to a feasible solution, a final NLP is solved in which the integer variables are fixed and the original objective is reinstalled on the continuous variables, similar to the sub-NLP heuristic described, for example, in [170].

Further, this paper considered solving a convex MINLP or a convex MIQP (mixed-integer quadratic programming problem) instead of a MIP in the rounding step, but gave computational evidence that this is not beneficial. To avoid cycling, their algorithm provides the MIP solver with a taboo list of previously used solutions. Linear constraints for the MIP problem are only generated from convex MINLP constraints. Finally, the authors showed that using an ℓ_∞-norm distance function instead of ℓ_1 as a MIP objective is competitive.

The work in [11] suggested three enhancements for an FP for nonconvex MINLP. First, the authors considered an automated selection of varying procedures for the rounding step. Second, they take into account the Hessian of the Lagrangian as part of the distance function to include second-order information while solving the rounding MIPs. Third, they dynamically generate linearization cuts for nonconvex constraints during the course of the FP algorithm.

7.2 LNS-Based Primal Heuristics

All primal heuristics presented in this section are obtained either by directly adapting an LNS-based heuristic originally introduced in the MIP context (see

Chapter 2) or by combining ingredients from those heuristics in various different ways.

For convex MINLP, the paper by Bonami and Gonçalves [38] introduced an extension of the RINS heuristic (see Section 2.2) to convex MINLPs. Recall that the idea of the RINS heuristic for MIPs is to combine two points that both fulfill partial requirements for an optimal MINLP solution: the LP relaxation solution is feasible and optimal for the linear constraints but typically not integer. The incumbent MIP solution is LP-feasible and integer but typically not optimal. Fixing variables that take equal values in both points maintains linear feasibility and creates a sub-MIP that still contains an LP-optimal point and the incumbent, and hopefully also contains improving integer solutions.

Bonami and Gonçalves use an optimum of the NLP relaxation as a second reference solution besides the incumbent. This follows the same idea as linear RINS: feasibility with respect to the constraints is maintained, and both reference points are part of the sub-MINLP. Interestingly, for their implementation in BONMIN, the resulting sub-MINLPs are solved by the Quesada and Grossmann algorithm [148], whereas for the original MINLP an NLP-based branch-and-bound algorithm is used. The former is much closer to the way most general-purpose global solvers, such as SCIP, BARON and FICO XPRESS, solve MINLPs (and sub-MINLPs). The authors employed a minimum fixing ratio of 10% and stop the sub-MINLP after the third improving solution, as has also been suggested in [16].

Concerning nonconvex MINLP, several methods use the same building blocks to either produce a first feasible solution or to improve a reference one. The methods discussed in the following differ in the way the explored large neighborhoods are defined, that is, either by *linear* or *nonlinear* constraints.

The iterative rounding heuristics introduced in [136] alternately solve NLPs and MIPs that are obtained by the relaxation of the original MINLP, local branching constraints (see Section 2.1) and restrictions of variable domains. These heuristics, named feasibility-based iterative rounding (FIR) and improvement-based iterative rounding (IIR), are developed to efficiently find feasible solutions or improve upon existing ones by iteratively rounding the solutions of continuous relaxations. Thereby, a couple of integer points close to the NLP solution are generated by solving an FP-like MIP and excluding the previously visited points by no-good cuts.

In [137], the authors suggested a local branching heuristic for nonconvex MINLPs. It solves a local branching MIP that is derived from a linear relaxation of the original MINLP, the integrality constraints and a local branching constraint (2.1). Notably, this constraint is only defined on binary variables, omitting general integers and thereby the need to introduce auxiliary variables and constraints. Another important aspect is how the objective function of the local branching MIP is defined, given that the MINLP objective is, in general, nonlinear and so can not be directly used for a MIP. The suggested idea is to solve the continuous nonlinear relaxation of the MINLP to local optimality and then using an ℓ_1-distance function to this local NLP optimum as the MIP objective. This resembles the way a distance function with respect to an integer point is used in heuristics such as the FP, proximity search or local branching repair; see Chapter 2.

If the local branching MIP finds a solution, this will be integer and feasible for the linear relaxation but not necessarily for the nonlinear constraints. To restore nonlinear feasibility, an NLP local search is performed by fixing the integer variables to the values from the local branching MIP's incumbent and solving the resulting continuous, nonconvex, nonlinear problem to local optimality.

This is a common practice in global outer-approximation-based MINLP solvers to deal with integer points that are feasible for the current linear relaxation of a MINLP. The integer variables get fixed, and the remaining continuous problem is solved to local optimality. Solvers like SCIP and FICO XPRESS differ in how they generate such integer candidate points, how frequently such a local search NLP is solved (for all or only for some solution candidates, immediately after generating them or after some waiting time) and the method to solve the NLP.

Unlike local branching for MIPs, this procedure might yield a point that is feasible and improving for the local branching MIP but infeasible or non-improving for the MINLP, after it has been adapted via the local NLP solve. The algorithm described in [137] overcomes this issue by executing the local branching procedure in a loop, adding a no-good constraint that forbids the previous local branching-MIP incumbent before taking the next iteration. In their experiments, they use a limit of at most 10 such iterations.

RECIPE, an acronym for *relaxed-exact continuous-integer problem exploration*, was introduced in [110, 111]. It is based on the variable neighborhood search paradigm, a meta-heuristic framework that does not work with a single predefined neighborhood as in large neighborhood search, but keeps changing the neighborhood definition, and often also its principle size, during the search. The neighborhoods in RECIPE are defined by local branching constraints and by restricting the bounds of general integer variables. Both the right-hand side of the local branching constraint and the considered interval lengths of the general integers dynamically change throughout the search, enlarging or shrinking the neighborhood as required. In this neighborhood, random points are sampled as starting points to find a local optimum of an NLP relaxation. Such a local optimum is then used as a reference point for a convex MINLP solver to find an integer point – as a solution candidate and as a reference point for the iterative neighborhood definition.

We end this section by mentioning a couple of methods that are not precisely LNS-based primal heuristics but are still somehow connected. Namely, meta-heuristic approaches, which are often used to compute promising reference points for multistart heuristics for nonconvex NLP and MINLP. As an example, consider the *OptQuest/NLP* algorithm [169]. HEXALY, formerly known as LOCALSOLVER [14, 70], is a commercial heuristic software that is based on local search. It combines several neighborhood search implementations with constraint propagation and MIP and NLP techniques.

7.3 Undercover

Undercover [23] is a large neighborhood search start heuristic that explores a sub-MIP derived from a given original MINLP. The algorithm is based on the insight that linearizing constraints through variable fixing can make nonlinear optimization problems much more tractable. Furthermore, each MINLP can be reduced to a sub-MIP by fixing sufficiently many variables. Undercover solves a vertex covering problem to identify a minimal set of variables to fix, a so-called *cover*, such that each nonlinear constraint becomes linear in the remaining variables. Formally, such a cover of an MINLP is defined as in Definition 7.2.

Definition 7.2 Let P be a MINLP of form (7.1) and $C \subseteq \mathcal{N}$ be a set of variable indices of P. We call C a *cover* of functions g_k, $k \in \mathcal{M}$ if and only if for all $\bar{x} \in [l, u]$, the set

$$\{(x, g_k(x)) : x \in [l, u], x_i = \bar{x}_i \text{ for all } i \in C\} \tag{7.3}$$

is an affine set intersected with $[l, u] \times \mathbb{R}$. We call C a *cover* of P if and only if C is a cover of all constraint functions $g_1, \ldots, g_m$.

To determine minimum covers, that is, minimal subsets of variables to fix in order to linearize each constraint of a MINLP, the authors suggested solving the vertex covering problem on the so-called *co-occurrence graph* of the MINLP [88].

Definition 7.3 (co-occurrence graph) Let P be a MINLP of the form (7.1) with $g_1, \ldots, g_m$ twice continuously differentiable on the interior of $[l, u]$. We call $G_P = (V_P, E_P)$ the *co-occurrence graph* of P with node set $V_P = \mathcal{N}$ given by the variable indices of P and edge set

$$E_P = \left\{ (i, j) \mid i, j \in V_P, \exists\, k \in \mathcal{M} : \frac{\partial^2}{\partial x_i \partial x_j} g_k(x) \not\equiv 0 \right\},$$

that is, an edge connects nodes i and j if and only if the Hessian matrix of some constraint has a structurally nonzero entry (i, j).

The undercover algorithm builds on the observation that, for a MINLP, a set $C \subseteq \mathcal{N}$ is a cover of P if and only if it is a vertex cover of the co-occurrence graph G_P. Berthold and Gleixner [24] showed that the majority of mixed-integer quadratically constrained programming problems (MIQCPs) from the GLoMIQO test set [131] and MINLPs from the MINLPLib [43] allow for covers consisting of at most 25% of the variables.

Once the cover is computed, the linearized sub-MIP is then constructed around a selected reference point, typically the optimal solution of an LP relaxation of the MINLP, aiming to preserve the feasibility of the solution with respect to the original nonlinear constraints. The idea is to strategically simplify the problem without fully discarding the nonlinear essence of the original problem. By solving a largest sub-MIP that still captures the crucial aspects of the MINLP, the heuristic can often find good-quality solutions.

Figure 7.2 illustrates this idea of undercover. The lightly shaded region shows the solid corresponding to the NLP relaxation; the parallel lines show a mixed-integer set of feasible solutions. The darkly shaded area shows the polyhedron associated with the undercover sub-MIP. The red circle B is the optimum of the NLP, the yellow circle A is the optimum of the undercover

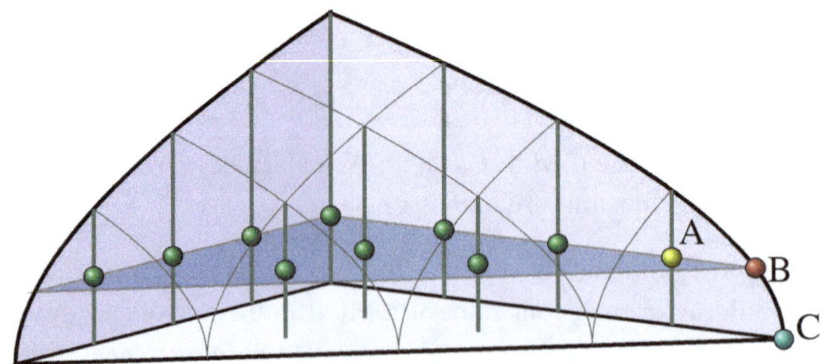

Figure 7.2 A convex MINLP and the undercover sub-MIP induced by the NLP relaxation. Used with permission of Springer Nature BC, from [23]; permission conveyed through Copyright Clearance Center, Inc.

sub-MIP, the light blue circle C is the optimum of the MIQCP. The green circles indicate further feasible solutions of the undercover sub-MIP.

As is typical for LNS heuristics, each feasible solution of the sub-MIP corresponds to a feasible solution of the original MINLP. Naturally, fixing variables to values obtained from a relaxation might render the MINLP infeasible. Undercover applies domain propagation to try to avoid infeasibilities and conflict analysis to learn additional constraints from infeasibilities that are nonetheless encountered.

Two major features distinguish undercover from other primal heuristics for MINLP that have been suggested in the literature. First, unlike most of them [24, 37, 38, 50, 137], undercover is not an extension of an existing MIP heuristic towards a broader class of problems. Moreover, it does not even have a clear counterpart in mixed-integer linear programming. Second, undercover solves two auxiliary MIPs (one for finding a set of variables to be fixed plus the resulting sub-MIP), and at most two NLPs (possibly one to compute initial fixing values and one for postprocessing the sub-MIP solution). In contrast to undercover, many heuristics [37, 50, 110, 111, 136, 137] for MINLP solve an arbitrarily large series of MIPs, often alternated with a sequence of NLPs, to produce a feasible start solution. The number of iterations is typically not fixed, but depends on the instance at hand.

The undercover heuristic's effectiveness is highlighted by its computational results, showing notable improvements in the performance of the SCIP solver, especially for MIQCPs, but also for other MINLPs. Its ability to complement existing root-node heuristics and provide significant benefits across a variety of problem instances marks it as a valuable addition to the solver's toolkit. The heuristic's general approach, requiring no specific problem structure beyond the ability to linearize constraints, makes it widely applicable and a promising area for further research and development in the field of optimization.

7.4 Quiz

It's quiz time! This section is designed to test your understanding of the concepts covered in this chapter. Each question is followed by four possible answers, only one of which is correct. If you are unsure about some questions, you might find it helpful to work through the corresponding section again. Remember, sometimes the process of elimination can be as valuable a tool as direct problem-solving. By engaging with this quiz, you'll reinforce your knowledge and identify areas requiring further study. Take your time, think carefully and check the answers in the Appendix when you are done.

Question 1 Which of the following is a typical characterization of MINLPs, one that has a huge impact on theoretical solvability and on the assumptions that algorithms can make?

☐ local and global MINLPs;
☐ convex and nonconvex MINLPs;
☐ heuristic and exact MINLPs;
☐ polynomial and exponential MINLPs.

Question 2 Which of these heuristics solves a MIP whose solutions are guaranteed to be feasible for the nonlinear constraints of the given MINLP?

☐ iterative rounding;
☐ undercover;
☐ nonlinear feasibility pump;
☐ RINS.

Question 3 A typical step in heuristics for MINLP is to fix integer variables and solve the resulting NLP

☐ to local optimality;
☐ to regional optimality;
☐ to global optimality;
☐ without an optimality guarantee.

Question 4 From a complexity point of view, solving the continuous relaxation of a nonconvex MINLP to global optimality is

☐ trivial;
☐ as hard as solving the continuous relaxation of a MIP;
☐ harder than solving the continuous relaxation of a MIP but not as hard as solving a MIP;
☐ harder than solving a MIP and as hard as solving a nonconvex MINLP.

8

Machine Learning for Primal Heuristics

The last decade has witnessed an impressive research effort around the use of machine learning (ML) in conjunction with many scientific disciplines. Roughly speaking, ML is a collection of techniques for learning patterns in or understanding the structure of data, often with the aim of performing data mining, that is, recovering previously unknown, actionable information from the learned data. Two main paradigms have emerged: *imitation learning* (IL) and *reinforcement learning* (RL). Typically, in IL one has to learn from data (points in the so-called training set) a (nonlinear) function that predicts a certain *score* for new data points that are not in the training set. Each data point is represented by a set of *features* that define its characteristics and whose patterns should be learned. Typically, a *label* is associated to each data point (or observation) in the training set. Here, the label stands for the score that the expert (a human or an algorithm) has assigned to the data point and that IL has to imitate. The techniques that are used to do IL are diverse, with *neural networks* (NNs) having recently been at the forefront. Conversely, RL follows the *Markov decision process* paradigm [98], where an agent sequentially makes decisions that have the effect of changing the state of the environment that represents the problem. For each decision, the agent collects a reward, and the goal is to learn a policy that maximizes the overall reward.[1]

This renewed interest in ML outside of the artificial intelligence community was triggered by the extraordinary success that modern statistical learning methods, especially *deep learning* (DL) [82], have obtained in areas such as image recognition, natural language processing and sequential games. Because

---

[1] A detailed description of ML, IL and RL, besides the few lines above, is outside of the scope of this book. The interested reader is referred, among other textbooks, to [33, 82, 167].

103

of those successes, it became natural to ask if those methods could be leveraged in other branches of science, and MIP was no exception to this trend.[2]

Of course, the relationship between ML and mathematical programming goes far beyond ML and MIP because convex optimization is a foundational building block of ML. Although there are notable exceptions (see, e.g., [162]), the ML relationship with combinatorial optimization (CO) is part of this latest investigation effort, and particularly relevant to this book is the work devoted to the use of ML for improving CO, an area often referred to as ML4CO [12]. It is important to stress, however, that the most classical way of integrating ML and CO is to use the former to estimate characteristics of a model through data and then solve the resulting (data-driven) CO model.

The most relevant amount of work in ML4CO has been devoted to improving branching [118] and, in particular, to the task of approximating strong branching [71, 84, 118], but the area of heuristics has also received a lot of attention. In this context, it is important to distinguish two different directions. The first follows the so-called *end-to-end learning* idea, that is, the use of ML to replace CO algorithms and directly predict the solution of a CO problem (the reader is referred to [107] for a recent survey covering the topic in detail). The second, which is extremely relevant for this book, concerns the use of modern statistical learning to improve primal heuristics for MIP (the reader is referred to [155] for a survey on the topic). In this chapter, we consider two specific areas of improvement. In Section 8.1, we discuss the use of ML in the deployment of primal heuristics, with special attention paid to scheduling them in SCIP [46], while in Section 8.2, we learn about some of the parameters or configurations of primal heuristics with a focus on learning for local branching [113]. Because the field is relatively recent, the chapter gives less of an overview and, instead, goes more into the details of some methods with respect to other chapters in the book. Section 8.3 ends the chapter with a short discussion on some perspectives on ML for primal heuristics.

8.1 Deploying Heuristics

As discussed in multiple chapters of the book, SCIP (as any other MIP solver) has a variety of primal heuristics to execute in the attempt to either find the first feasible solution or improve the current incumbent. In Chapter 6, we discussed how different primal heuristics have different performances, and their success rate depends on several factors. First, the performance of a specific heuristic in

---

[2] An even more recent spark of interest has been generated by the successes of *large language models*, but their use within MIP has not been systematically explored so far.

the arsenal of any MIP solver cannot be considered in isolation. This is because several of these primal heuristics exploit variations of a similar idea, so their chances of success often depend on the *order of execution*. Second, the success of any heuristic h_1 should be evaluated not only with respect to the number of times in which it finds an improving solution, but also with respect to the *amount of time* it takes to execute it because, as just discussed, maybe another heuristic h_2 that is scheduled to be executed after h_1 could have found the same solution in a fraction of the computing time. In other words, an effective use of the primal heuristics in a MIP solver would require establishing the best order in which to execute them and, based on this order, for how long each heuristic should be executed.

8.1.1 Scheduling Heuristics in SCIP

Traditionally, the order of execution of a set of k heuristics $\mathcal{H} = \{h_1, h_2, \ldots, h_k\}$ is established with an informal statistical process that executes some sensible configurations on a specific dataset and selects the best one according to classical MIP metrics, largely unrelated to the considerations above. Moreover, the amount of time allowed to each h_i, $i = 1, \ldots, k$, of the same type (i.e., in the same class like LNS heuristics) is roughly the same with respect to some deterministic effort limits (e.g., LP iterations or branch-and-bound nodes).

In [46], the joint task of deciding the execution order in $\mathcal{H}$ and the time allowed for each $h_i \in \mathcal{H}$ was instead cast formally as a *scheduling problem* and solved in a data-driven way. Namely, at each branch-and-bound node d in which heuristics should be executed, and for each heuristic, $h_i \in \mathcal{H}$, $t(h_i, d)$ denotes the number of iterations[3] required by h_i to find a feasible solution, where $t(h_i, d) = +\infty$ if h_i cannot find it. Then, in a totally offline manner, the scheduling problem can be formulated as a mixed-integer quadratic program (MIQP), where the overall number of iterations is minimized under the policy that the sequence of heuristics executions is aborted as soon as a heuristic is successful in finding an improving solution.

Solving the MIQP, even in an offline fashion, is impractical not only because it would require knowing $t(h_i, d)$ (which can be estimated, see below), but mainly because the number of branch-and-bound nodes in which heuristics are executed is very large, especially for problems where the scheduling makes a significant difference. Instead, the framework developed in [46] solves

---

[3] The number of iterations is used as a deterministic proxy of the computing time. The definition of what an iteration is depends on the type of heuristic. For example, the maximal diving depth is used for diving heuristics, while branch-and-bound nodes or simplex iterations can be used for sub-MIP heuristics.

the problem in a greedy way after an extensive data collection aimed at learning $t(h_i, d)$.

More precisely, SCIP is executed in a so-called shadow mode, that is, each of the heuristics is run extensively, and its performance recorded (success or not, and if successful, in how many iterations), but if a solution is found, it is not passed to the solver to avoid having the order in which the primal heuristics are executed influence the success rate. This strategy is very effective and allows for unbiased and extensive data collection, although the success rate recorded is not with respect to the improvement of the incumbent solution but only with respect to finding feasible solutions. This data collection is able to accurately record the $t(h_i, d)$ values, and the scheduling problem is solved greedily[4] by following this simple policy inductively: for each possible pair (h_i, τ), where h_i is a heuristic that has not yet been scheduled and τ is a possible number of iterations in which the heuristic could be executed (estimated by $t(h_i, d)$), the pair that maximizes the ratio between the number of additional nodes d in which a feasible solution is found and the time spent (i.e., iterations) is selected and added to the schedule.

This scheme provides a flexible mechanism for IL in the scheduling heuristic context. Indeed, the $t(h_i, d)$'s can be interpreted as labels that are collected by running the heuristics independently and measuring their success. The optimization problem is solved in a greedy fashion, but it is not exactly trying to match those labels as is customary in many IL approaches. In fact, in IL the objective function of the associated optimization problem (the so-called loss) is generally the distance between the predicted labels and the observed ones. Here, there is no loss function, but the process is analogous because (i) the labels are observed during the data collection, (ii) the optimization problem is solved offline, and (iii) the scheduling obtained offline is applied at inference time, that is, when a new instance has to be solved. It is also easy to see that the measure of success can be cumulated, that is, we can safely record $t(h_i, d)$'s for many MIP instances; the more data, the better according to the classical ML mantra, especially in the DL context. However, the merging of data coming from different sources, (from different classes of MIP instances in the MIP context) does not come for free. It is conceivable that MIP instances with different structures could admit different schedules both in terms of the order and the number of iterations. In other words, it is natural that, for example, a primal heuristic showing great performance for set covering MIP models could be less effective for combinatorial auctions MIP models.

This observation highlights what, so far, has been a clear limitation of ML4CO: in almost all cases, an ML model developed to discover a pattern

---

[4] The greedy scheme is an adaptation of the one developed in [166] in an online context.

in data coming from a MIP solving context[5] has to be trained with the data coming from a specific distribution of the MIP instances. In other words, generalization outside of the distribution in ML4CO has been almost nonexistent. In the specific context at hand, the ML model for the primal heuristics scheduling in [46] was trained, separately, on two families of MIP instances, namely, the generalized independent set and the fixed-charge multicommodity network flow problems, and the results obtained show a significant improvement with respect to the SCIP default scheduling order (and amount of effort, i.e., iterations). We will come back to this issue in Section 8.3.

8.1.2 More Scheduling Tasks

In [92], an adaptive large neighborhood search was implemented in SCIP to decide the sequence in which eight LNS primal heuristics should be executed. Such a sequence is learned through a multiarmed bandit approach in an online fashion by applying an effective compromise between intensification (i.e., continue running a heuristic that has proven to be effective) and diversification (i.e., try different primal heuristics) to extend the overall capability of the framework.

From a scheduling perspective, the question answered in [46] is not the only relevant one. Indeed, probably the most basic question is whether primal heuristics should be executed in a branch-and-bound node or not. This is the question considered in [104], which was in the early days of the use of ML for MIP. Note that this decision is not trivial to make even in an ideal context in which one knows if a heuristic will be effective because the overall runtime of an instance might not be directly affected in a positive way (the evolution of the branch and bound changes unpredictably). Still, the work in [104] compared an approach that runs a heuristic when successful with an optimal offline schedule and then imitates it, obtaining good results with respect to the reduction of the primal integral.

8.2 Learning Parameters

Primal heuristics rely on a number of parameters that control their deployment and determine their capability of delivering good solutions fast. Traditionally, these parameters are set without using statistical learning methods. Sections 8.2.1 and 8.2.2 discuss attempts in the direction of using ML for that.

---

[5] The MIP solving context here is intended as the area of MIP solving one wants to apply ML to. Besides scheduling primal heuristics, some of the contexts explored so far in the literature have been variable selection in branching, parameter setting in preprocessing and cutting plane selection, just to mention a few; see [12].

8.2.1 Revisiting Local Branching through an ML Lens

The local branching scheme as introduced in Section 2.1 was surprisingly effective even if the choice of its parameters was not done in a sophisticated statistical way. The first of those parameters is the size of the (large) neighborhood that the MIP solver explores. In [62], the value k in the local branching constraint (2.1) was rather arbitrarily fixed to 20.[6] It is easy to see that such an initial choice, especially in the context of the LB implementation as a primal heuristic within a MIP solver, can have a severe effect on the effectiveness of the method. Indeed, the LB integration as a primal heuristic in MIP generally comes with a very limited number of consecutive neighborhood explorations, depending on the fact that improvements are obtained at each call. In the most crude version, independently of the success, only a single neighborhood is explored every time the incumbent solution is updated. Thus, an initial, poor choice of k can make the method potentially ineffective. In [113], ML was used to predict a reasonable value for k for a specific class of instances. Such prediction also depends on the initial solution LB is applied to, that is, the very first solution found by the MIP solver or that at the end of the root node exploration. What the prediction is looking at is the "sweet spot" for the neighborhood size k where the quality is high, that is, the incumbent solution can be improved significantly and the exploration is fast (so the computing time to explore the neighborhood is small). In [113], it was shown empirically that such a sweet spot exists; see Figure 8.1.

The data collection for the prediction task is easier to set up than the one discussed in Section 8.1. Indeed, it suffices to explore any (distinct) neighborhood size $k = \lceil r \times N \rceil$, where $N = \mathcal{B}$ is the number of binary variables, $r \in [0, 1]$, and record the quality of the solution and the time it takes to explore the neighborhood (by potentially applying a time limit). Those recordings are used as labels within a regression task that leverages both static and dynamic features collected during the LB procedure, for example, context of the problem, incumbent solution, solving status and computational cost. The ML model used for the regression task is a *graph neural network* (GNN) that has proven to be useful in several other ML4CO contexts [44, 45]. Indeed, GNNs are able to capture the relevant structure of the MIP constraint matrix by representing it as a bipartite graph with nodes, on the one side associated with variables, and on the other side, associated with constraints: a link between the two nodes in different sets, that is, an edge in the graph, exists when a variable has a nonzero

---

[6] A second version of constraint (2.1) was explored in [62], where one pays attention only to the support of $\tilde{x}$, that is, only to the variables $\tilde{x}_j$, $j \in \mathcal{B}$ such that $\tilde{x}_j = 1$. This is referred to as the *asymmetric* version of constraint (2.1) and when used in [62] the initial value of k was set to 10.

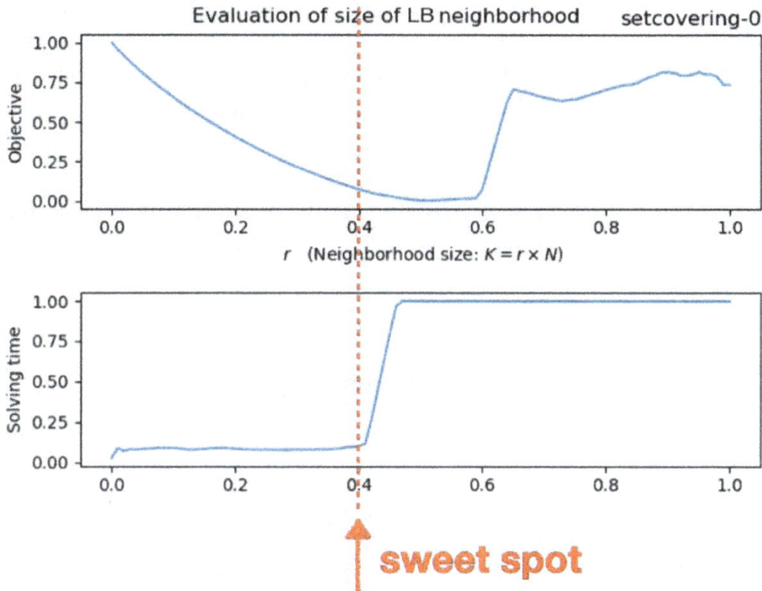

Figure 8.1 Evaluation of the size of the LB neighborhood on a set covering instance (setcovering-0): $k = \lceil r \times N \rceil$ (where $N = |\mathcal{B}|$ is the number of binary variables, $r \in [0, 1]$ (slightly different notation with respect to the figure). A time limit is imposed for each neighborhood exploration. Image courtesy of Defeng Liu.

coefficient in a constraint. This network is clearly able to capture the static features (matrix structure, degree of the nodes as density, etc.), while the dynamic ones are associated as vectors to either nodes or edges and change over time while the GNN is used for prediction. As such, a different GNN was trained in [113] for each class of MIP instances considered, and it was shown that doing that gives non-negligible benefits.

As briefly mentioned in Section 2.1, originally the local branching framework as presented in [62] embedded the exploration of the neighborhood defined by constraint (2.1) in an iterative approach that (i) applies a new local branching constraint each time the incumbent solution has been improved (after removing the previous LB constraint), and (ii) dynamically adapts k based on the effectiveness of the neighborhood exploration. In a framework like this, the initial value of k is slightly less crucial because it is dynamically adapted. However, once again, the adaptive scheme in [62] was very simple and, interestingly, it can be significantly improved by ML [113, 114]. Specifically, an RL approach was used to learn a policy that adapts both the size of the neighborhood k and the computing time that should be given to each neighborhood exploration. Remarkably, it was shown in [114] that if LB is embedded within an adaptive scheme, one can avoid using a GNN trained on a specific class of MIP instances it to predict the initial k. More precisely, the algorithm of [114] recovers almost the same results by training the GNN on an heterogeneous pool of 0-1 MIP instances, that is, instances where the structure of the constraint matrix captured by the GNN varies. It would probably be too much to say that this algorithm achieves generalization outside of the distribution because the class considered is that of the 0-1 MIP instances, but this is still a significant and partially unexpected result.

8.2.2 More ML-Guided LNS Heuristics

Several other LNS approaches have been designed in the literature by exploiting ML predictions. In [55], ML was used to predict a subset of the binary variables that had high probability of belonging to an optimal solution, and a local branching constraint is applied to the subset. In the same vein, [105, 135] used the prediction to do LNS by hard-fixing the predicted variables and solving the resulting sub-MIP. Interestingly, in [105] the predicted variables were also used to guide node selection in branching. Finally, all the papers mentioned in Section 8.2.2 cast the prediction problem as an IL task by using feasible solutions, not necessarily optimal ones, as initial labels.[7]

8.3 Perspectives

The examples discussed in this chapter are rather encouraging. They show that significant gains can be achieved by using modern statistical learning to leverage data coming from the solution of MIP instances. This is a good achievement because the first few ML4CO years were devoted to proving that learning in CO was possible, but not much attention was given to showing that improved performances could be achieved. The work considered here, as well as that in other MIP contexts (see, e.g., [22, 26, 39, 71, 168]), shows that the gains are with respect to state-of-the-art algorithms. However, there are at least two major issues that have slowed down the implementation of ML-based algorithmic tools in MIP.

The first challenge is that discussed at the end of Section 8.1.1, that is, the difficulty of generalizing the ML models outside of the distribution in which they are trained. This issue somehow collides with the standard MIP paradigm

---

[7] More work in this context was reviewed in [155].

that prescribes the development of MIP solvers working within a default configuration, that is, an algorithmic setting that should be good enough for any MIP instance. This paradigm has allowed the impressive development of MIP technology because it has forced scientists and software developers to think about general methods and to select those ideas that proved to be most robust. This enabled MIP to be considered a black box technology, and it has certainly been beneficial at a time where the technology was rapidly developing and the room for improvement was wide. However, this opportunity has shrunk now that MIP technology has become more mature, and it might be that the next generation of (strong) improvements could be achieved mainly by some form of specialization, which could be associated with the modern statistical learning tools that ML provides that are able to discover patterns that inherently depend on classes of instances. The picture provided at the end of Section 8.2.1, where some form of generalization is achieved across classes of MIP instances, indicates that it is not yet clear if specialization for MIP development will be a requirement. Nevertheless, the results achieved so far by training and testing in the same distribution for MIP tasks are not generally achievable if training and testing are performed on heterogenous classes of MIP instances.

The second challenge is more technological. Many of the more effective ML models, namely those using NNs (e.g., for GNN in [113] or [71]), are better implemented on graphical processing units (GPUs) whose main characteristic is to perfectly exploit parallelism. However, GPUs are not the computing architecture that MIP solvers commonly use.[8] In fact, MIP solvers are developed and run on central processing units (CPUs), and making GPUs and CPUs coexist, so as to take advantage of both in the best possible way, is currently a technological open problem. The development of hybrid architectures currently seems to be the most viable direction, and this idea was explored in [84], where a GNN was used parsimoniously at the root node of a MIP and is then replaced for the rest of the computation (i.e., in the branch-and-bound nodes) by other more CPU-friendly ML models for efficient variable selection in branching (on CPU machines).

8.4 Quiz

It's quiz time! This section is designed to test your understanding of the concepts covered in this chapter. Each question is followed by four possible answers, only one of which is correct. If you are unsure about some questions,

---

[8] On the contrary, so far the attempts to use GPUs for MIP technology have been rather disappointing.

you might find it helpful to work through the corresponding section again. Remember, sometimes the process of elimination can be as valuable a tool as direct problem-solving. By engaging with this quiz, you'll reinforce your knowledge and identify areas requiring further study. Take your time, think carefully and check the answers in the Appendix when you are done.

Question 1 Scheduling primal heuristics in a MIP solver involves deciding at each node

- ☐ the order in which they are executed;
- ☐ if any of them has to be executed;
- ☐ for how long any of them is executed;
- ☐ all of the above.

Question 2 A good generalization in machine learning for MIP is achieved when one can successfully use an ML model on

- ☐ instances larger than those the model has been trained on;
- ☐ instances belonging to the training set;
- ☐ instances obtained by combining those in the training set;
- ☐ the continuous relaxation of the instances.

Question 3 A graph neural network captures the structure of a MIP by

- ☐ associating variables to values;
- ☐ representing it as a bipartite graph;
- ☐ using attention mechanisms in a graph;
- ☐ representing it as a multigraph.

Question 4 The size of the neighborhood of a local branching constraint

- ☐ can be learnt by taking the MIP class into account;
- ☐ depends on the number of continuous variables;
- ☐ is equal to 18;
- ☐ is directly proportional to the number of variables.

Appendix

Quiz Solutions

Solutions for Chapter 1

Answer 1 *always*. A finite number of linear constraints always defines a polyhedron.

Answer 2 *all variables are up-locked*. There is only one constraint, of $\leq$-type, and all coefficients are positive.

Answer 3 *is monotonically decreasing and reaches zero when the algorithm terminates*. The global primal bound can only improve over time (as we find better and better incumbents), and the global dual bound, obtained as the best dual bound among the open nodes, can also only improve over time, as the primal-dual gap shrinks monotonically. Finally, the algorithm is an exact method for integer programming, and thus the primal-dual gap is zero on termination.

Answer 4 *practically relevant and effective at reducing the primal gap*. While it is true that branch and bound does not need primal heuristics for convergence, the availability of a good primal solution early in the search improves the practical performance of the method and significantly reduces the primal integral. The impact on the time to prove optimality is much less pronounced but still notable.

Answer 5 *an incomplete but polynomial inference scheme*. Constraint propagation is not a complete method. However, if each variable is binary, then at least it can be run in polynomial time (propagating a single linear constraint can be done in polynomial time, and at each round, at least one variable gets fixed, or the procedure stops).

Answer 6 *has no guarantees*. Despite being a very good starting point when designing a heuristic, by itself the approach is not guaranteed to find a feasible solution.

Answer 7 *is a measure balancing solution quality and runtime.* The primal integral depends on the time limit unless the optimal solution is found before that. Also, it is not normalized to $[0, 1]$ unless we divide by the time limit itself.

Solutions for Chapter 2

Answer 1 *the neighborhoods are defined in a problem-independent way, exploiting the MIP technology.* Most MIP LNS heuristics do diversify (usually relying on the global branch-and-bound search) and some have a randomized component (e.g., solution polishing). It is not true that the resulting sub-MIP is too expensive to solve to optimality (although it can be, hence the strict resource limits).

Answer 2 *restricts the search to solutions within distance k with respect to the ℓ_1-norm.* The local branching can be stated for general integer variables and it still restricts the solution space resulting in a proper sub-MIP, albeit not by hard-fixing variables.

Answer 3 *is always feasible.* It contains by construction the current incumbent.

Answer 4 *can be run even without an incumbent.* RENS is a starting heuristic, not an improving heuristic. Still, it does need at least an LP solution, and it is not guaranteed to be feasible (there might be no feasible rounding).

Answer 5 *extends the RINS logic in its combination step.* The mutation step can be run even with only a single feasible solution in the pool, and it does not systematically pick the best two solutions in the pool to increase diversification. Finally, because of its strong randomization component, it does not need an underlying branch-and-bound search to diversify its behavior.

Solutions for Chapter 3

Answer 1 *might not give a feasible solution.* Rounding, in general, gives no guarantees whatsoever.

Answer 2 *random/against the objective.* In a binary knapsack problem, the continuous relaxation has at most one fractional variable (the one corresponding to the so-called *critical item*). Rounding toward the objective or to the nearest integer, regardless of the variable order, can yield a solution exceeding the capacity of the knapsack. On the other hand, rounding against the objective will always set the critical item to zero, and keep the other variables the same, thus giving a feasible solution.

Answer 3 *all of them, even without backtracking.* Constraint propagation will take care of fixing to zero any item that no longer fits into the container, so fix-and-propagate will find a feasible solution no matter the variable order and to which value each variable is fixed.

Answer 4 *so as to not spoil the search with branching decisions unsuited for the optimality proof.* Branch and bound, being a complete algorithm, will eventually find an optimal solution even without any primal heuristic. Also, no heuristic is guaranteed to improve performance, and diving heuristics are certainly not without overhead, exactly because they need to solve additional LPs.

Solutions for Chapter 4

Answer 1 *one is LP feasible and one is integer feasible.* The LP feasible point is the outcome of the projection step, while the integer feasible point is obtained via the rounding step.

Answer 2 *dealing with cycles.* While the other points are relevant as well, the most important aspect is to reduce the occurrence of cycles and how to deal with them when they (invariably) occur, as that affects the performance of the whole scheme.

Answer 3 *a WalkSAT-like scheme.* Without any further modifications to the original FP scheme, only the scheme in [54] guarantees convergence.

Solutions for Chapter 5

Answer 1 *the original basic variable goes to one of its bounds.* Pivoting out of the basis of an original basic variable implies putting it at one of its bounds, that is, it becomes integral.

Answer 2 *yes.* The idea of line search is very general and allows maximum flexibility with respect to the initial point(s).

Answer 3 *it has favorable geometric properties.* The intuition for using the analytic center, especially for problems with general integer variables, is that of exploiting its geometric position in the interior of the LP relaxation.

Solutions for Chapter 6

Answer 1 *improving the primal bound early in the search.* Primal heuristics have a modest impact on time to optimality and only an indirect impact on the dual bound. For proving infeasibility, they are typically detrimental since they do not contribute toward that goal. It is getting good solutions early in the search where heuristics shine – and what makes them so important in practice.

Answer 2 *a shifted geometric mean.* Large numbers dominate arithmetic means, and small numbers dominate geometric means. A shifted geometric mean balances this effect – in an arithmetic mean, a shift would simply cancel out.

Answer 3 *the mix of all types of heuristics.* Taking a single heuristic, or even a class of heuristics, out of the picture does not often have a drastic effect because other heuristics can compensate, at least partially. When running without any heuristics, performance clearly degrades in all measures.

Answer 4 *almost doubles.* For individual instances or homogeneous instance sets, results can, of course, differ drastically. Still, for heterogeneous instance sets, this number is in line with what other studies with other instance sets and other solvers found earlier.

Solutions for Chapter 7

Answer 1 *convex and nonconvex MINLPs.* The main advantage of convex MINLPs is that their continuous relaxation can be solved (up to a certain tolerance) in polynomial time. Also, for convex functions and feasible regions, local optimality implies global optimality. The distinction of local versus global, therefore, is made to characterize solutions, not problems, and heuristic versus exact is a characterization of algorithms.

Answer 2 *undercover.* Undercover creates a MIP from a MINLP by fixing variables. Consequently, the MIP will be a subproblem of the MINLP and if it has feasible solutions, they will also be feasible for the original MINLP. Iterative rounding and the feasibility pump solve MIP relaxations, whereas RINS solves a sub-MINLP.

Answer 3 *to local optimality.* Regional optimality is not a common concept. For nonconvex problems, solving the NLP to global optimality could be as hard as solving the original MINLP. Local optimality (up to certain tolerances) is achievable in polynomial time under some mild assumptions. Local branching, undercover, and some feasibility pump variants solve an

NLP to local optimality to find feasible (and locally optimal) values for the continuous variables of a MINLP when fixing the integer variables to values that are feasible for a linear relaxation of the MINLP.

Answer 4 *harder than solving a MIP and as hard as solving a nonconvex MINLP.* LP-solving is possible in polynomial time; MIP-solving, MINLP-solving and solving nonconvex NLP(s) are NP-hard. But while MIP is in NP, both MINLP-solving and solving nonconvex NLP(s) are not; they are, theoretically, undecidable problems.

Solutions for Chapter 8

Answer 1 *all of the above.* Scheduling primal heuristics for a MIP solver is a very complex task that involves many decisions. Not all of them have been considered simultaneously in the literature.

Answer 2 *instances larger than those the model has been trained on.* Currently, ML technology is successful when, within the same distribution, one can generalize to larger instances. Generalizing outside of the distribution is still hard to achieve.

Answer 3 *representing it as a bipartite graph.* The bipartite graph has variables on one side and constraints on the other side. A connection is established if a variable is present in a constraint.

Answer 4 *can be learnt by taking the MIP class into account.* The ML model in [113] is able to learn an effective value for the initial neighborhood size by training a graph neural network. Such value gives a good compromise between speed in the exploration of the neighborhood and quality of the solution obtained.

References

[1] T. Achterberg. *Constraint integer programming*. PhD thesis, Technische Universität Berlin, 2007.

[2] T. Achterberg. SCIP: Solving constraint integer programs. *Mathematical Programming Computation*, 1(1):1–41, 2009.

[3] T. Achterberg. LP basis selection and cutting planes. Presentation slides from MIP 2010 workshop in Atlanta. www.isye.gatech.edu/sites/default/files/documents/conferences/2010mip/mip-program.pdf, 2010.

[4] T. Achterberg. LP relaxation modification and cut selection in a MIP solver. US patent, US20110131167, 2011.

[5] T. Achterberg and T. Berthold. Improving the feasibility pump. *Discrete Optimization*, 4(1):77–86, 2007.

[6] T. Achterberg and R. Wunderling. Mixed integer programming: Analyzing 12 years of progress. In M. Jünger and G. Reinelt, editors, *Facets of Combinatorial Optimization – Festschrift for Martin Grötschel*, pages 449–481. Springer, 2013.

[7] R. K. Ahuja, J. B. Orlin, and D. Sharma. Very large-scale neighborhood search. *International Transactions in Operational Research*, 7(4–5):301–317, 2000.

[8] D. Baena and J. Castro. Using the analytic center in the feasibility pump. *Operations Research Letters*, 39(5):310–317, 2011.

[9] E. Balas, S. Ceria, M. Dawande, F. Margot, and G. Pataki. OCTANE: A new heuristic for pure 0-1 programs. *Operations Research*, 49(2):207–225, 2001.

[10] E. Balas and C. H. Martin. Pivot and complemen: a heuristic for 0–1 programming. *Management Science*, 26(1):86–96, 1980.

[11] P. Belotti and T. Berthold. Three ideas for a feasibility pump for nonconvex MINLP. *Optimization Letters*, 11(1):3–15, 2017.

[12] Y. Bengio, A. Lodi, and A. Prouvost. Machine learning for combinatorial optimization: A methodological tour dhorizon. *European Journal of Operational Research*, 290(2):405–421, 2021.

[13] M. Bénichou, J.-M. Gauthier, P. Girodet, G. Hentges, G. Ribière, and O. Vincent. Experiments in mixed-integer programming. *Mathematical Programming*, 1:76–94, 1971.

[14] T. Benoist, B. Estellon, F. Gardi, R. Megel, and K. Nouioua. LocalSolver 1.x: A black-box local-search solver for 0–1 programming. *4OR – A Quarterly Journal of Operations Research*, 9(3):299–316, 2011.

[15] L. Bertacco, M. Fischetti, and A. Lodi. A feasibility pump heuristic for general mixed-integer problems. *Discrete Optimization*, 4(1):63–76, 2007.

[16] T. Berthold. *Primal heuristics for mixed integer programs.* Diploma thesis, Technische Universität Berlin, 2006.

[17] T. Berthold. Measuring the impact of primal heuristics. *Operations Research Letters*, 41(6):611–614, 2013.

[18] T. Berthold. *Heuristic algorithms in global MINLP solvers.* PhD thesis, Technische Universität Berlin, 2014.

[19] T. Berthold. RENS: the optimal rounding. *Mathematical Programming Computation*, 6(1):33–54, 2014.

[20] T. Berthold. Finding Proper Hardware to Improve your Optimization Software Performance. `https://community.fico.com/s/blog-post/a5Q4w000000CvM9EAK/fico2741`, 2021.

[21] T. Berthold, T. Feydy, and P. J. Stuckey. Rapid learning for binary programs. In A. Lodi, M. Milano, and P. Toth, editors, *Proceedings of CPAIOR 2010*, volume 6140 of *Lecture Notes in Computer Science*, pages 51–55. Springer, June 2010.

[22] T. Berthold, M. Francobaldi, and G. Hendel. Learning to use local cuts. *arXiv preprint arXiv:2206.11618*, 2022.

[23] T. Berthold and A. M. Gleixner. Undercover: A primal MINLP heuristic exploring a largest sub-MIP. *Mathematical Programming*, 144(1–2):315–346, 2014.

[24] T. Berthold, S. Heinz, M. E. Pfetsch, and S. Vigerske. Large neighborhood search beyond MIP. In L. D. Gaspero, A. Schaerf, and T. Stützle, editors, *Proceedings of the 9th Metaheuristics International Conference (MIC 2011)*, pages 51–60, 2011.

[25] T. Berthold and G. Hendel. Shift-and-propagate. *Journal of Heuristics*, 21(1):73–106, 2015.

[26] T. Berthold and G. Hendel. Learning to scale mixed-integer programs. In *Proceedings of the AAAI Conference on Artificial Intelligence*, volume 35, pages 3661–3668, AAAI, 2021.

[27] T. Berthold, G. Hendel, and T. Koch. From feasibility to improvement to proof: Three phases of solving mixed-integer programs. *Optimization Methods and Software*, 33(3):499–517, 2018.

[28] T. Berthold, A. Lodi, and D. Salvagnin. Ten years of feasibility pump, and counting. *EURO Journal on Computational Optimization*, 7(1):1-14, 2019.

[29] T. Berthold, M. Perregaard, and C. Mészáros. Four good reasons to use an interior point solver within a MIP solver. In N. Kliewer, J. Ehmke, and R. Borndörfer, editors, *Operations Research Proceedings 2017*, pages 159–164, Springer, 2018.

[30] T. Berthold and D. Salvagnin. Cloud branching. In C. Gomes and M. Sellmann, editors, *Integration of AI and OR Techniques in Constraint Programming for Combinatorial Optimization Problems*, volume 7874 of *Lecture Notes in Computer Science*, pages 28–43. Springer, 2013.

[31] T. Berthold, P. J. Stuckey, and J. Witzig. Local rapid learning for integer programs. In L.-M. Rousseau and K. Stergiou, editors, *Integration of Constraint Programming, Artificial Intelligence, and Operations Research*, pages 67–83. Springer, 2019.

[32] K. Bestuzheva, M. Besançon, W.-K. Chen, et al. The SCIP Optimization Suite 8.0. *arXiv preprint arXiv:2112.08872*, 2021.

[33] C. M. Bishop. *Pattern Recognition and Machine Learning. Information Science and Statistics*. Springer, 2006.

[34] N. L. Boland, A. C. Eberhard, F. G. Engineer et al. Boosting the feasibility pump. *Mathematical Programming Computation*, 6:255–279, 2014.

[35] N. L. Boland, A. C. Eberhard, F. G. Engineer, and A. Tsoukalas. A new approach to the feasibility pump in mixed integer programming. *SIAM Journal on Optimization*, 22(3):831–861, 2012.

[36] P. Bonami, L. T. Biegler, A. R. Conn, et al. An algorithmic framework for convex mixed integer nonlinear programs. *Discrete Optimization*, 5:186–204, 2008.

[37] P. Bonami, G. Cornuéjols, A. Lodi, and F. Margot. A feasibility pump for mixed integer nonlinear programs. *Mathematical Programming*, 119(2):331–352, 2009.

[38] P. Bonami and J. Gonçalves. Heuristics for convex mixed integer nonlinear programs. *Computational Optimization and Applications*, 51:729–747, 2012.

[39] P. Bonami, A. Lodi, and G. Zarpellon. A classifier to decide on the linearization of mixed-integer quadratic problems in CPLEX. *Operations Research*, 70(6):3303–3320, 2022.

[40] I. Borosh and L. B. Treybig. Bounds on positive integral solutions of linear Diophantine equations. *Proceedings of the American Mathematical Society*, 55(2):299–304, 1976.

[41] A. Brearley, G. Mitra, and H. Williams. Analysis of mathematical programming problems prior to applying the simplex algorithm. *Mathematical Programming*, 8:54–83, 1975.

[42] R. Burkard, M. DellAmico, and S. Martello. *Assignment Problems*. SIAM, 2009.

[43] M. R. Bussieck, A. S. Drud, and A. Meeraus. MINLPLib – a collection of test models for mixed-integer nonlinear programming. *INFORMS Journal on Computing*, 15(1):114–119, 2003.

[44] Q. Cappart, D. Chételat, E. B. Khalil et al. Combinatorial optimization and reasoning with graph neural networks. In *IJCAI*, pages 4348–4355, 2021.

[45] Q. Cappart, D. Chételat, E. B. Khalil et al. Combinatorial optimization and reasoning with graph neural networks. *Journal of Machine Learning Research*, 24:1–61, 2023.

[46] A. Chmiela, E. Khalil, A. Gleixner, A. Lodi, and S. Pokutta. Learning to schedule heuristics in branch and bound. *Advances in Neural Information Processing Systems*, 34:24235–24246, 2021.

[47] P. M. Christophel. *An improved heuristic for the MOPS mixed integer programming solver*. Diploma thesis, Universität Paderborn, 2005.

[48] T. H. Cormen, C. E. Leiserson, R. L. Rivest, and C. Stein. *Introduction to Algorithms*. MIT Press, 2001.

[49] C. D'Ambrosio, A. Frangioni, L. Liberti, and A. Lodi. Experiments with a feasibility pump approach for nonconvex MINLPs. In P. Festa, editor, *Experimental Algorithms*, volume 6049 of *Lecture Notes in Computer Science*, pages 350–360. Springer, 2010.

[50] C. D'Ambrosio, A. Frangioni, L. Liberti, and A. Lodi. A storm of feasibility pumps for nonconvex MINLP. *Mathematical Programming*, 136:375–402, 2012.

[51] E. Danna, E. Rothberg, and C. L. Pape. Exploring relaxation induced neighborhoods to improve MIP solutions. *Mathematical Programming*, 102(1):71–90, 2004.

[52] M. De Santis, S. Lucidi, and F. Rinaldi. A new class of functions for measuring solution integrality in the feasibility pump approach. *SIAM Journal on Optimization*, 23(3):1575–1606, 2013.

[53] M. De Santis, S. Lucidi, and F. Rinaldi. Feasibility pump-like heuristics for mixed integer problems. *Discrete Applied Mathematics*, 165:152–167, 2014.

[54] S. S. Dey, A. Iroume, M. Molinaro, and D. Salvagnin. Improving the randomization step in feasibility pump. *SIAM Journal on Optimization*, 28(1):355–378, 2018.

[55] J.-Y. Ding, C. Zhang, L. Shen et al. Accelerating primal solution findings for mixed integer programs based on solution prediction. In *Proceedings of the AAAI Conference on Artificial Intelligence*, volume 34, pages 1452–1459, 2020.

[56] M. A. Duran and I. E. Grossmann. An outer-approximation algorithm for a class of mixed-integer nonlinear programs. *Mathematical Programming*, 36(3):307–339, 1986.

[57] J. Eckstein and M. Nediak. Pivot, cut, and dive: A heuristic for 0–1 mixed integer programming. *Journal of Heuristics*, 13(5):471–503, 2007.

[58] B. H. Faaland and F. S. Hillier. Interior path methods for heuristic integer programming procedures. *Operations Research*, 27:1069–1087, 1979.

[59] T. A. Feo and M. G. C. Resende. A probabilistic heuristic for a computationally difficult set covering problem. *Operations Research Letters*, 8:67–71, 1989.

[60] FICO Xpress Optimizer. `www.fico.com/en/products/fico-xpress-optimization`.

[61] M. Fischetti, F. Glover, and A. Lodi. The feasibility pump. *Mathematical Programming*, 104(1):91–104, 2005.

[62] M. Fischetti and A. Lodi. Local branching. *Mathematical Programming*, 98(1–3):23–47, 2003.

[63] M. Fischetti and A. Lodi. Repairing MIP infeasibility through local branching. *Computers & Operations Research*, 35(5):1436–1445, 2008.

[64] M. Fischetti and M. Monaci. Branching on nonchimerical fractionalities. *Operations Research Letters*, 40(3):159–164, 2012.

[65] M. Fischetti and M. Monaci. Proximity search for 0-1 mixed-integer convex programming. *Journal of Heuristics*, 20(6):709–731, 2014.

[66] M. Fischetti and D. Salvagnin. Feasibility pump 2.0. *Mathematical Programming Computation*, 1:201–222, 2009.

[67] L. R. Ford and D. R. Fulkerson. Maximal flow through a network. *Canadian Journal of Mathematics*, 8(3):399–404, 1956.

[68] M. Frank and P. Wolfe. An algorithm for quadratic programming. *Naval Research Logistics Quarterly*, 3(1–2):95–110, 1956.

[69] G. Gamrath, T. Berthold, S. Heinz, and M. Winkler. Structure-driven fix-and-propagate heuristics for mixed integer programming. *Mathematical Programming Computation*, 11:675–702, 2019.

[70] F. Gardi. *Toward a mathematical programming solver based on local search.* Habilitation thesis, Université Pierre et Marie Curie, 2013.

[71] M. Gasse, D. Chételat, N. Ferroni, L. Charlin, and A. Lodi. Exact combinatorial optimization with graph convolutional neural networks. In *Proceedings of the 33rd International Conference on Neural Information Processing Systems*, pages 15580–15592. Curran Associates, 2019.

[72] J.-M. Gauthier and G. Ribière. Experiments in mixed-integer linear programming using pseudo-costs. *Mathematical Programming*, 12(1):26–47, 1977.

[73] B. Geißler, A. Morsi, L. Schewe, and M. Schmidt. Penalty alternating direction methods for mixed-integer optimization: A new view on feasibility pumps. *SIAM Journal on Optimization*, 27(3):1611–1636, 2017.

[74] S. Ghosh. DINS, a MIP improvement heuristic. In M. Fischetti and D. P. Williamson, editors, *Integer Programming and Combinatorial Optimization, 12th International IPCO Conference, Proceedings*, volume 4513 of *Lecture Notes in Computer Science*, pages 310–323. Springer, 2007.

[75] S. Ghosh and R. B. Hayward. Pivot and Gomory cut: A MIP feasibility heuristic. Technical report, University of Alberta, 2010.

[76] A. Gleixner, L. Gottwald, and A. Hoen. Papilo: A parallel presolving library for integer and linear programming with multiprecision support. *INFORMS Journal on Computing*, 35(6):1329–1341, 2022.

[77] F. Glover. Tabu search – part I. *ORSA Journal on Computing*, 1:190–206, 1989.

[78] F. W. Glover and G. A. Kochenberger, editors. *Handbook of Metaheuristics*, volume 57 of *International Series in Operations Research & Management Science*. Kluwer / Springer, 2003.

[79] R. E. Gomory. Outline of an algorithm for integer solutions to linear programs. *Bulletin of the American Mathematical Society*, 64(5):275–278, 1958.

[80] R. E. Gomory. An algorithm for the mixed integer problem. Technical report, RAND Corporation, 1960.

[81] J. Gondzio. Presolve analysis of linear programs prior to applying an interior point method. *INFORMS Journal on Computing*, 9:73–91, 1997.

[82] I. Goodfellow, Y. Bengio, and A. Courville. *Deep Learning*. MIT Press, 2016.

[83] J. E. Graver. On the foundations of linear and integer linear programming I. *Mathematical Programming*, 9(1):207–226, 1975.

[84] P. Gupta, M. Gasse, E. Khalil, et al. Hybrid models for learning to branch. *Advances in Neural Information Processing Systems*, 33:18087–18097, 2020.

[85] Gurobi optimization. www.gurobi.com/.

[86] K. Halbig, A. Hoen, A. Gleixner, J. Witzig, and D. Weninger. A diving heuristic for mixed-integer problems with unbounded semi-continuous variables. *arXiv preprint arXiv:2403.19411*, 2024.

[87] S. Hanafi, J. Lazić, and N. Mladenović. Variable neighbourhood pump heuristic for 0-1 mixed integer programming feasibility. *Electronic Notes in Discrete Mathematics*, 36(0):759–766, 2010.

[88] P. Hansen and B. Jaumard. Reduction of indefinite quadratic programs to bilinear programs. *Journal of Global Optimization*, 2(1):4160, 1992.

[89] R. M. Haralick and G. L. Elliott. Increasing tree search efficiency for constraint satisfaction problems. *Artificial Intelligence*, 14(3):263–313, 1980.

[90] U.-U. Haus, M. Köppe, and R. Weismantel. The integral basis method for integer programming. *Mathematical Methods of Operations Research*, 53(3):353–361, 2001.

[91] G. Hendel. *Empirical analysis of solving phases in mixed integer programming.* Master's thesis, Technische Universität Berlin, 2014.

[92] G. Hendel. Adaptive large neighborhood search for mixed integer programming. *Mathematical Programming Computation*, 14(2):185–221, 2022.

[93] G. Hendel, M. Miltenberger, and J. Witzig. Adaptive algorithmic behavior for solving mixed integer programs using bandit algorithms. In B. Fortz and M. Labbé, editors, *Operations Research Proceedings 2018*, pages 513–519, 2019.

[94] F. S. Hillier. Efficient heuristic procedures for integer linear programming with an interior. *Operations Research*, 17:600–637, 1969.

[95] A. J. Hoffman, M. Mannos, D. Sokolowsky, and N. A. Wiegmann. Computational experience in solving linear programs. *Journal of the Society for Industrial and Applied Mathematics*, 1(1):17–33, 1953.

[96] K. L. Hoffman and M. Padberg. Solving airline crew scheduling problems by branch-and-cut. *Management Science*, 39(6):657–776, 1993.

[97] J. N. Hooker. *Integrated Methods for Optimization*. Springer, 2007.

[98] R. A. Howard. *Dynamic Programming and Markov Processes*. MIT Press, 1960.

[99] T. Huang, A. Ferber, Y. Tian, B. Dilkina, and B. Steiner. Local branching relaxation heuristics for integer linear programs. In A. A. Cire, editor, *International Conference on the Integration of Constraint Programming, Artificial Intelligence, and Operations Research*, pages 96–113. Springer, 2023.

[100] T. Ibaraki, T. Ohashi, and H. Mine. A heuristic algorithm for mixed-integer programming problems. *Mathematical Programming Study*, 2:115–136, 1974.

[101] IBM ILOG CPLEX Optimizer. www-01.ibm.com/software/integration/optimization/cplex-optimizer/.

[102] T. Junttila and P. Kaski. Engineering an efficient canonical labeling tool for large and sparse graphs. In D. Applegate, G. S. Brodal, D. Panario, and R. Sedgewick, editors, *Proceedings of the Ninth Workshop on Algorithm Engineering and Experiments and the Fourth Workshop on Analytic Algorithms and Combinatorics*, pages 135–149. SIAM, 2007.

[103] R. Karp. Reducibility among combinatorial problems. In R. Miller and J. Thatcher, editors, *Complexity of Computer Computations*, pages 85–103. Plenum Press, 1972.

[104] E. B. Khalil, B. Dilkina, G. L. Nemhauser, S. Ahmed, and Y. Shao. Learning to run heuristics in tree search. In *Proceedings of the Twenty-sixth International Joint Conference on Artificial Intelligence*, pages 659–666, 2017.

[105] E. B. Khalil, C. Morris, and A. Lodi. MIP-GNN: A data-driven framework for guiding combinatorial solvers. In L.-M. Rousseau and K. Stergiou, editors, *Proceedings of the AAAI Conference on Artificial Intelligence*, volume 36, pages 10219–10227, 2022.

[106] T. Koch, T. Achterberg, E. Andersen et al. MIPLIB 2010. *Mathematical Programming Computation*, 3(2):103–163, 2011.

[107] J. Kotary, F. Fioretto, P. Van Hentenryck, and B. Wilder. End-to-end constrained optimization learning: A survey. In Z.-H. Zho, editor, *Proceedings of the Thirtieth International Joint Conference on Artificial Intelligence*, pages 4475–4482, 2021.

[108] A. H. Land and A. G. Doig. An automatic method of solving discrete programming problems. *Econometrica*, 28(3):497–520, 1960.

[109] A. N. Letchford and A. Lodi. Primal cutting plane algorithms revisited. *Mathematical Methods of Operations Research*, 56(1):67–81, 2002.

[110] L. Liberti, N. Mladenović, and G. Nannicini. A recipe for finding good solutions to MINLPs. *Mathematical Programming Computation*, 3:349–390, 2011.

[111] L. Liberti, G. Nannicini, and N. Mladenović. A good recipe for solving MINLPs. In V. Maniezzo, T. Stützle, and S. Voß, editors, *Matheuristics*, volume 10 of *Annals of Information Systems*, pages 231–244. Springer, 2010.

[112] P. Lin, S. Cai, M. Zou, and J. Lin. New characterizations and efficient local search for general integer linear programming. *arXiv preprint arXiv:2305.00188*, 2024.

[113] D. Liu, M. Fischetti, and A. Lodi. Learning to search in local branching. In K. Sycara, V. Honaver, and M. Spaan, editors, *Proceedings of the AAAI Conference on Artificial Intelligence*, 36(4):3796–3803, 2022.

[114] D. Liu, M. Fischetti, and A. Lodi. Revisiting local branching with a machine learning lens. *arXiv preprint arXiv:2112.02195*, 2022.

[115] Localsolver: News. `www.localsolver.com/news.html?id=32`. Accessed July 30, 2014.

[116] A. Lodi. Mixed integer programming computation. In M. Jünger, T. M. Liebling, D. Naddef, G. L. Nemhauser, W. R. Pulleyblank, G. Reinelt, G. Rinaldi, and L. A. Wolsey, editors, *50 Years of Integer Programming 1958–2008*, pages 619–645. Springer, 2010.

[117] A. Lodi and A. Tramontani. Performance variability in mixed-integer programming. In *Theory Driven by Influential Applications*, pages 1–12. INFORMS, 2013.

[118] A. Lodi and G. Zarpellon. On learning and branching: A survey. *TOP*, 25:207–236, 2017.

[119] A. Løkketangen, K. Jörnsten, and S. Storøy. Tabu search within a pivot and complement framework. *International Transactions in Operational Research*, 1(3):305–316, 1994.

[120] H. R. Lourenço, O. C. Martin, and T. Stützle. Iterated local search: Framework and applications. In F. Glover and G. Kochenberger, editors, *Handbook of Metaheuristics*, volume 57, pages 321–353. Kluwer Academic Publishers, 2002.

[121] B. Luteberget and G. Sartor. Feasibility jump: An LP-free Lagrangian MIP heuristic. *Mathematical Programming Computation*, 15:365–388, 2023.

[122] H. Mahmoud and J. W. Chinneck. Achieving MILP feasibility quickly using general disjunctions. *Computers & Operations Research*, 40(8):2094–2102, 2013.

[123] V. Maniezzo, M. A. Boschetti, and T. Stützle. *Matheuristics, Algorithms and Implementations*. Springer, 2021.

[124] I. Maros. *Computational Techniques of the Simplex Method*. Kluwer Academic Publishers, 2002.

[125] S. Martello and P. Toth. *Knapsack Problems: Algorithms and Computer Implementations*. Wiley, 1990.

[126] A. Martin. *Integer programs with block structure*. Habilitation thesis, Preprint SC 99-03, Zuse Institute Berlin, 1999.

[127] G. Mexi, T. Berthold, and D. Salvagnin. Using multiple reference vectors and objective scaling in the feasibility pump. *EURO Journal on Computational Optimization*, 11:1–18, 2023.

[128] G. Mexi, M. Besançon, S. Bolusani et al. Scylla: A matrix-free fix-propagate-and-project heuristic for mixed-integer optimization. *arXiv preprint arXiv:2307.03466*, 2023.

[129] Miplib2017. `http://miplib.zib.de`.

[130] R. Misener and C. A. Floudas. Global optimization of mixed-integer quadratically-constrained quadratic programs (MIQCQP) through piecewise-linear and edge-concave relaxations. *Mathematical Programming*, 136:155–182, 2012.

[131] R. Misener and C. A. Floudas. GloMIQO: Global mixed-integer quadratic optimizer. *Journal of Global Optimization*, 57:3–50, 2013.

[132] M. Mitchell. *An Introduction to Genetic Algorithms*. MIT Press, 1996.

[133] M. H. Moskewicz, C. F. Madigan, Y. Zhao, L. Zhang, and S. Malik. Chaff: Engineering an efficient SAT solver. In *Proceedings of the 38th Design Automation Conference*, pages 530–535. ACM, 2001.

[134] L.-M. Munguía, S. Ahmed, D. A. Bader, G. L. Nemhauser, and Y. Shao. Alternating criteria search: A parallel large neighborhood search algorithm for mixed integer programs. *Computational Optimization and Applications*, 69:1–24, 2018.

[135] V. Nair, S. Bartunov, F. Gimeno et al. Solving mixed integer programs using neural networks. *arXiv preprint arXiv:2012.13349*, 2020.

[136] G. Nannicini and P. Belotti. Rounding-based heuristics for nonconvex MINLPs. *Mathematical Programming Computation*, 4(1):1–31, 2012.

[137] G. Nannicini, P. Belotti, and L. Liberti. A local branching heuristic for MINLPs. *arXiv preprint arXiv:0812.2188*, 2008.

[138] J. Naoum-Sawaya. Recursive central rounding for mixed integer programs. *Computers & Operations Research*, 43:191–200, 2014.

[139] J. Naoum-Sawaya and S. Elhedhli. An interior point cutting plane heuristic for mixed integer programming. *Computers & Operations Research*, 38:1335–1341, 2011.

[140] G. L. Nemhauser and L. A. Wolsey. *Integer and Combinatorial Optimization*. Wiley, 1988.

[141] G. L. Nemhauser, L. A. Wolsey, and M. L. Fisher. An analysis of approximations for maximizing submodular set functions – I. *Mathematical Programming*, 14(1):265–294, 1978.

[142] C. Neumann, S. Schwarze, O. Stein, and B. Müller. Feasible rounding based diving strategies in branch-and-bound methods for mixed-integer optimization. *EURO Journal on Computational Optimization*, 10:100051, 2022.

[143] C. Neumann, O. Stein, and N. Sudermann-Merx. A feasible rounding approach for mixed-integer optimization problems. *Computational Optimization and Applications*, 72:309–337, 2019.

[144] J. Ostrowski, J. Linderoth, F. Rossi, and S. Smriglio. Orbital branching. *Mathematical Programming*, 126(1):147–178, 2011.

[145] A. Pal and H. Charkhgard. A feasibility pump and local search based heuristic for bi-objective pure integer linear programming. *INFORMS Journal on Computing*, 31(1):115–133, 2019.

[146] J. Patel and J. W. Chinneck. Active-constraint variable ordering for faster feasibility of mixed integer linear programs. *Mathematical Programming*, 110:445–474, 2007.

[147] J. Pryor and J. W. Chinneck. Faster integer-feasibility in mixed-integer linear programs by branching to force change. *Computers & Operations Research*, 38(8):1143–1152, 2011.

[148] I. Quesada and I. E. Grossmann. An LP/NLP based branch and bound algorithm for convex MINLP optimization problems. *Computers & Chemical Engineering*, 16(10):937–947, 1992.

[149] F. Rossi, P. van Beek, and T. Walsh, editors. *The Handbook of Constraint Programming*. Elsevier, 2006.

[150] E. Rothberg. An evolutionary algorithm for polishing mixed integer programming solutions. *INFORMS Journal on Computing*, 19(4):534–541, 2007.

[151] R. M. Saltzman and F. S. Hillier. A heuristic ceiling point algorithm for general integer linear programming. *Management Science*, 38(2):263–283, 1992.

[152] D. Salvagnin. Detecting and exploiting permutation structures in MIPS. In H. Simonis, editor, *Proceedings of CPAIOR 2014*, volume 8451 of *Lecture Notes in Computer Science*, pages 29–44. Springer, 2014.

[153] D. Salvagnin, R. Roberti, and M. Fischetti. A fix-propagate-repair heuristic for mixed integer programming. Technical report, DEI, University of Padova, 2023.

[154] M. W. P. Savelsbergh. Preprocessing and probing techniques for mixed integer programming problems. *ORSA Journal on Computing*, 6:445–454, 1994.

[155] L. Scavuzzo, K. Aardal, A. Lodi, and N. Yorke-Smith. Machine learning augmented branch and bound for mixed integer linear programming. *arXiv preprint arXiv:2402.05501*, 2024.

[156] U. Schöning. A probabilistic algorithm for k-SAT and constraint satisfaction problems. In *Proceedings of 40th Annual Symposium on Foundations of Computer Science*, pages 410–414. IEEE, 1999.

[157] C. Schulte. *Programming constraint services*. PhD thesis, Universität des Saarlandes, Germany, 2000.

[158] C. Schulte and P. J. Stuckey. Speeding up constraint propagation. In M. Wallace, editor, *Principles and Practice of Constraint Programming – CP 2004, 10th International Conference*, volume 3258 of *Lecture Notes in Computer Science*, pages 619–633. Springer, 2004.

[159] S. Sharma. Mixed-integer nonlinear programming heuristics applied to a shale gas production optimization problem. Master's thesis, Norwegian University of Science and Technology, 2013.

[160] S. Sharma, B. R. Knudsen, and B. Grimstad. Towards an objective feasibility pump for convex MINLPS. *Computational Optimization and Applications*, 63:737–753, 2016.

[161] P. Shaw. Using constraint programming and local search methods to solve vehicle routing problems. In M. Maher and J.-F. Puget, editors, *Principles and Practice of Constraint Programming – CP98*, volume 1520 of *Lecture Notes in Computer Science*, pages 417–431. Springer, 1998.

[162] K. A. Smith. Neural networks for combinatorial optimization: A review of more than a decade of research. *INFORMS Journal on Computing*, 11:15–34, 1999.

[163] G. Sonnevend. An "analytical centre" for polyhedrons and new classes of global algorithms for linear (smooth, convex) programming. In A. Prékopa, J. Szelezsáan, and B. Strazicky, editors, *System Modelling and Optimization*, volume 84 of *Lecture Notes in Control and Information Sciences*, pages 866–875. Springer, 1986.

[164] G. Sonnevend. Applications of the notion of analytic center in approximation (estimation) problems. *Journal of Computational and Applied Mathematics*, 28:349–358, 1989.

[165] SoPlex. An open source LP solver implementing the revised simplex algorithm. `http://soplex.zib.de/`.

[166] M. Streeter. *Using online algorithms to solve NP-hard problems more efficiently in practice*. PhD thesis, Carnegie Mellon University, 2007.

[167] R. S. Sutton and A. G. Barto. *Reinforcement Learning: An Introduction*. MIT Press, 1998.

[168] M. Turner, T. Berthold, M. Besanon, and T. Koch. Cutting plane selection with analytic centers and multiregression. In *Proceedings of CPAIOR 2023*, volume 13884, pages 52–68, 2023.

[169] Z. Ugray, L. Lasdon, J. Plummer et al. Scatter search and local NLP solvers: A multistart framework for global optimization. *INFORMS Journal on Computing*, 19(3):328–340, 2007.

[170] S. Vigerske. *Decomposition in multistage stochastic programming and a constraint integer programming approach to mixed-integer nonlinear programming*. PhD thesis, Humboldt-Universität zu Berlin, 2012.

[171] C. Wallace. ZI round, a MIP rounding heuristic. *Journal of Heuristics*, 16(5):715–722, 2010.

[172] R. Weismantel. Test sets of integer programs. *Mathematical Methods of Operations Research*, 47(1):1–37, 1998.

[173] J. Witzig and A. Gleixner. Conflict-driven heuristics for mixed integer programming. *INFORMS Journal on Computing*, 33(2):706–720, 2021.

[174] R. Wunderling. *Paralleler und Objektorientierter Simplex-Algorithmus*. PhD thesis, Technische Universität Berlin, 1996.

[175] R. D. Young. A simplified primal (all-integer) integer programming algorithm. *Operations Research*, 16(4):750–782, 1968.

Index

For EU product safety concerns, contact us at Calle de José Abascal, 56–1°,
28003 Madrid, Spain or eugpsr@cambridge.org.